HISTORIC WINTER STORMS OF NEW JERSEY

HISTORIC WINTER STORMS OF NEW JERSEY

DON COLGAN

Published by The History Press
An imprint of Arcadia Publishing
Charleston, SC
www.historypress.com

First published 2025

Manufactured in the United States

ISBN 9781467170000

Library of Congress Control Number: 2025941491

Notice: The information in this book is true and complete to the best of our knowledge. It is offered without guarantee on the part of the author or The History Press. The author and The History Press disclaim all liability in connection with the use of this book.

CONTENTS

FOREWORD

Every meteorologist starts off as a weather enthusiast. For as long as I can remember, I was fascinated with weather. Weather enthusiasts are able to see nature in a different way. They come to marvel at the best of nature, as well as the awe and power that nature flexes from time to time. Some do grow up to be professionals in the field, and others take different paths. However, the enthusiasm for weather goes with them for the rest of their lives. In that lifetime, there will be countless snowstorms, hurricanes, nor'easters and other severe weather events. They will remember every one of them, as Don Colgan has done. He puts his experiences into writing to remind everyone of storms past and their impacts. I was more than happy to be part of his process, bringing together his experiences with prior storms. It is a way of keeping a record of weather history from a personal viewpoint, and it keeps the magic of weather alive in the hearts of every weather freak out there!

Joe Cioffi, meteorologist

PREFACE

I can remember the moment my lifelong love affair with winter storms began. It was December 12, 1961. I was living at 120 Locust Street in Roselle in a three-room upstairs apartment where our back porch was the roof. I was in kindergarten, it was Monday and I was to be escorted by my mother the five blocks to Locust Elementary School. I looked out the window and the snow had not only covered the cars parked in front, but it also reached six feet up the telephone pole, where it had been thrown by the plows. We had twenty-one inches of snow. It was an extraordinary moment for me, frozen in time in my mind. My curiosity and interest in winter storms was born.

To this day—and I am seventy years of age—I consider myself a weatherman. I never did realize my ambition to become a meteorologist. It was my childhood dream, but life circumstances kept that dream just beyond my grasp. My parents moved to Elizabeth in 1962, and that is where I spent my boyhood. I have catalogued just about every noteworthy snowstorm in my mind, from 1964 until 2024, ready for instant recall. My father was a truck driver, and while my mother worked in Cranford Hall as a nurse's aide, she was largely a stay-at-home mom, helping me with my homework and signing my report cards, which contained mixed results at best.

I attended Marquis De Lafayette Junior High School in the North End of Elizabeth and went to Thomas Jefferson High School, where I graduated in 1973. I had no hope of going to college; my father had passed away earlier in 1973, and I had to supplement my seven-day-a-week job delivering newspapers for Academy Newspaper Service with a full-time job as a typist clerk at Lindquist Steel near the Elizabeth Port.

Growing up, I moved a lot, and my interest in meteorology grew, compounded by my love of snow and winter storms. I played baseball and touch football with my friends, the "North End gang," at Kellogg Park, a big park and playground that was the "go to" place for kids in the North End.

Every night, at 11:00 p.m., I would turn on Channels 4 and 7 and endure the newscast for those precious few minutes when Dr. Frank Field or Tex Antoine would give the weather report. Tex was pure entertainment, and he was usually right. Dr. Field captured your attention; he was decades ahead of his time, not only forecasting but also explaining the developing weather situation. I remember Tex telling viewers on the night of Saturday, February 8, 1969, "It may get dicey north of 135th Street. I'll be back Monday with apologies." Meanwhile, every forecaster was calling for snow that would quickly change to rain early Sunday. Apologize he did; it was the infamous Lindsay snowstorm that dropped an equally surprising fifteen to twenty inches of snow and paralyzed the city.

With transistor radio in hand, waiting for the 5:00 p.m. revised National Weather Bureau forecast and dropping dimes into any available phone booth to get the latest forecast, I was married for life to weather, specifically winter storms. My fifth-grade teacher, Mr. Ruzinski, at Aldene Elementary School in Roselle Park, where I lived for one year, referred to me as "weatherman, par excellence." My seventh-grade history teacher at Theodore Roosevelt Junior High school in Elizabeth, Mr. Primiano, more than once told me to "get your head out of the window and into your textbook."

When I was twelve years old, I borrowed Howard J. Critchfield's *General Climatology* from the Elizabeth Public Library and read it cover to cover. A wonderfully instructive book, it was the beginning of my self-education in meteorology. One of my earliest mentors in weather and a fellow lover of winter storms was my cousin Richard, a graduate of Rutgers University. We followed the weather together and talked on the phone about what the chances of snow were for Elizabeth. He taught me to believe in myself, bolstering my confidence that I had a future as a writer and a weatherman.

I remember Christmas night 1969, as an early winter storm, a nor'easter, was moving up the Eastern Seaboard, promising to bring a last-minute white Christmas. Snow was in the forecast, up to a half foot, before a forecast changed to rain the following morning. Richie came bursting into the living room, "Donnie, they just issued heavy snow warnings, six to eight inches." It was my last Christmas present of 1969.

There is an anticipation, a tension and a thrill when a winter storm arrives. Snowstorms are rarely a certainty. For an all-snow event in New Jersey, from Cape May to High Point, ample cold air has to be in place and low pressure has to ride along what is called the "40–70 benchmark," which is far enough offshore to lock in the cold air with a northeasterly wind component. The warmer Atlantic waters are always lurking nearby, threatening to turn a forecast of snowfall into puddles and busy windshield wipers. The storm comes closer to the shoreline, and the winds turn southeasterly, becoming the killer of snowflakes.

Then there is the rain-snow line, that sliver of real estate that separates heavy rain from busy snowplows. There are nor'easters that ride just inside the benchmark, creating a battle zone between rain, sleet and snow. There was a strong coastal storm on February 19, 1972, that forecasters could not get a handle on. Snow was originally forecast, with six to ten inches predicted for New York City and Northern New Jersey. Tex Antoine disappointed snow lovers at eleven o'clock Saturday night with the latest information from "the machines." The coastal low was then expected to hug the coastline, bringing in warmer air, with snow changing rapidly to an all-day washout Sunday.

The storm couldn't make up its mind. It stayed farther offshore, and heavy snow developed after midnight and accumulated three inches before daybreak. Then it changed to a wind-driven mixture of snow, sleet and rain. The rain-snow line jumped around all day, with rain changing to snow and then back to rain. Finally, cold air wrapped around the departing low-pressure area, and precipitation changed to snow before ending, adding another two inches of accumulation. It was an exciting storm!

I've found that people often forget snowstorms yet love to discuss the ones that are entrenched in their memories. Once, while I was discussing weather with a close friend, he said, "Don, it doesn't snow any more in Jersey; it's always rain." I had to remind him that since 2000, we have had four blizzards and at least eight snowstorms that dropped more than twelve inches of snow, four of them with twenty inches of accumulation. The period from 2000 to 2021 was extremely active in terms of jet stream phasing, resulting in powerful winter storms. During the 2010s, there were four consecutive winters with over forty inches of snow recorded at Newark Liberty Airport.

There have been so many giants in meteorology. They guided us from childhood to, in some instances, our senior years. They were trailblazers we grew to trust, a fraternity of weather forecasting excellence—Tex Antoine,

Dr. Frank Field, Gordon Barnes, Bob Harris, Nick Gregory, Bill Korbel, Jim Witt, Stuart Soroka, Craig Allen, Joe Cioffi, Joe Rao, the iconic Alan Kasper, John Coleman, Ira Joe Fisher, Dan Zarrow, Janice Huff, Dr. Joel Myers, Elliot Abrams and so many more.

ACKNOWLEDGEMENTS

Historic Winter Storms of New Jersey, a book centered on my sixty-year passion for meteorology and the science of weather forecasting, has been a life goal. There are people who, in the course of my lifetime, provided support and encouragement and, most importantly, helped me believe in myself.

My mother, who sat with me and helped me with my English homework; my father, who worked until he died at seventy-three, never enjoying the fruits of retirement. My cousin Richard, who kindled my interest in weather and told me on Christmas night in 1969, as a nor'easter and heavy snow was bearing down on New Jersey, "Donnie, you are intelligent. You can be a writer."

My very first employer, Simon Bornstein, who owned a small newspaper company, Academy Newspaper Service, and paid me five dollars a week to deliver papers in the North End of Elizabeth. He worked seven days a week and drilled his work ethic into me. He was stern yet impeccably honest. When my father died, he told me that afternoon, "Donald, you're a man now. You have to take care of your mother."

My fifth-grade teacher at Aldene Elementary School in Roselle Park, New Jersey, Robert Ruzinski, who was the first to recognize my love of weather and called me "the classroom weatherman, par excellence."

My lifelong best friend Dave Ryan. He has been at my side for a lifetime—from sandlot baseball at Kellogg's Park in Elizabeth to being my best man at my wedding, through seven decades of cold beer, baseball, boxing and fatherhood—with humor, a world of knowledge and a passion for his St. Louis Cardinals.

My boss and mentor Tom Knott, who nurtured my career in electrical distribution, built and strengthened my creative skills and gave me support in a most difficult time in my life. He believed in me.

Finally, my dear wife, Kathy; my daughter, Courtney; and my son, James, the rocks in my life, gave me needed criticism, encouragement, patience and support. They helped me wade through the challenges of my lifelong goal of becoming an author.

I would like to recognize those who helped me with this project. Their patience, mentoring and willingness to help made this book—my lifelong ambition—a reality. Meteorologist Joe Cioffi answered every question I sent him, usually within five minutes, and provided the foreword for this book. I would also like to thank meteorologist Joe Rao, a respected forecaster with a deep background in the field of astronomy, and New Jersey State Climatologist David Robinson. Meteorologist Jim Witt, with generations of experience and unsurpassed knowledge, is a teacher and innovator, a creator, who provided me with his narrative and photos of the December 26, 1947 Boxing Day snowstorm that he experienced as a boy.

Jim, a pioneer in the science of long-range outlooks, is the creator of Weather Wiz, which provides long-range daily forecasts, several years out, for any location in the United States. Craig Allen, who worked seven days a week on WCBS Radio and weekend television, took the time to tell me his remembrances of the Blizzard of '96 and the Storm of the Century in March 1993. Meteorologists Joe Martucci and Kyle David shared their family remembrances of the great Christmas weekend Blizzard of 2010.

I could never forget Margaret Thomas Buchholz, an acclaimed Jersey Shore author who helped guide me through the Great Atlantic Storm of 1962.

The Sea Isle City Historical Society, with storm survivors Joe LaRosa and Pat Haffert, took me back to the devastating March 1962 gale, also called the "Ash Wednesday storm," and relived that catastrophic and destructive nor'easter, a harrowing storm that brought enormous destruction and loss of life to the New Jersey barrier islands.

The Cape May Historical Society, with president and historian Harry Bellangy and administrator and secretary Kathleen C. Wyatt, walked me through the Great Atlantic Storm of 1962 and the Christmas weekend Blizzard of 2010. They provided images and YouTube clips of the historic storms and took me through each one of the five high tides that demolished much of the New Jersey shoreline, particularly those of the vulnerable barrier islands.

Jack Osborn, who spent his boyhood summers in the historic sand beach community that bore his name, Camp Osborn, also deserves my thanks. Over eighty cottages and homes, which had survived blizzards, tropical storms and hurricanes, were destroyed in a Sandy-inspired fire on October 29, 2012. Jack patiently told me the story of the Camp's history, back to the depression years in the early 1930s. Camp Osborn is the heart, soul and fabric of the New Jersey Shore.

This book has been a journey, from Cape May to High Point State Park, with New Jerseyans from all walks of life—salesmen, truck drivers, businessmen and women, housewives, public officials, meteorologists, writers and teachers—each with a story to tell.

I'd also like to thank the following:

Al Mungo
Alan Kasper (meteorologist)
Alan Schutz
Allison McQuillan
Betty Ann Fuller
Bill Hoffman
Bob Gathanhy
Bob Ziff
Boro of Lavallette
Caitlin Haffert
Christopher C. Burt
Christopher Stacheleski (meteorologist)
Courtney Elizabeth Colgan
Craig Allen (meteorologist)
Craig Benner
Dan Valle
Dan Zarrow (meteorologist)
Daryl Susan Mitruska-Syms
Dave Dabour
Dave Ryan
David Robinson (New Jersey state climatologist)
David Zimmer
Ed Miro
Emil R. Salbini
Eric Thomas, meteorologist for WBTV (Charlotte, NC)
Erik Larsen
Extreme Weather Watch
Frank Infantino
Frank Tursi
George Friedman
The Greater Cape May Historical Society, Sulke family collection
Harry Bellangy
IBS World
Jack Osborn
James Patrick Colgan
Jean Mickle
Jeff Goldman
Jill Pharo
Jim Witt (meteorologist)
Joe Cioffi (meteorologist)
Joe LaRosa
Joe Martucci (meteorologist)
Joe Rao (meteorologist)
John Esselin

Kaitlin Haffert
Kathleen C. Wyatt
Kathleen Gaffney Colgan
Ken Pileggi
Kyle David (meteorologist)
Larry Savadove
Lauren M. Black
Louis Uccellini
Margaret Thomas Bucholtz
Marion Osborn
Mark Leberfinger
Michael Estrin
Michelle Smayda
Mike and Sue Minutella
Moshe Kinderlehrer
National Weather Service (NOAA.gov)
NBC Philadelphia
Newark Star Ledger staff
New Jersey Weather Observers (NJWO)
Nick Gregory (meteorologist)
Nick Stefano
Nimbus Earth Observatory
NJ.com, New Jersey Advanced Media
NJ 101.5
Pat Haffert
Paul C. Kocin
Paul Mickle
Raymond G. Fisk
Ryan Grenoble
Sally Goldenberg
Scott Mazella
Sea Isle City Historical Society
Sergio Bichao
Stacy Geisenger
Stan Kozinski
Stephen Stirling
Steve Dedinsky
Sussex County Historical Society
Tony Monardo
United States Department of Commerce
UPI Archives
Wayne Blanchard
Wayne Smith
Wayne T. McCabe
Weather Underground
WHYY

Introduction

A WINTER STORM'S JOURNEY

It happens every time a winter storm threatens New Jersey. Another meteorologist critic—and they are legion—spouts, "They are wrong 90 percent of the time and get to keep their jobs. If that was me on my job, I'd be fired."

The exact opposite is true. Forecast accuracy is over 99 percent, and when you consider the complexity and difficult decisions forecasters have to make, receiving myriad data from global and short- and long-range computer models, you should be thankful that you have your own occupation and not that of a meteorologist.

The following is the development and evolution of a typical New Jersey winter storm. Now, you are the meteorologist. How would you interpret the data and issue a forecast?

An extratropical wave of low pressure entered the United States mainland over the Pacific Northwest on Tuesday, January 17, bringing rain to Seattle and other coastal regions and snow to the mountains. The global models ECMWF, GFS and UKMO (UKMET), which forecast for up to two weeks out, begin signaling that this storm system will drive southeastward, into the lower Midwest and then into the Tennessee valley. Snow occurs in most locations along the path of the storm, with rain and embedded thunderstorms affecting southern Tennessee, Alabama and Georgia. There was remarkable consistency all week that indicated the southern low would transfer its energy to a primary low-pressure center off the South Carolina coastline and deepen rapidly into a nor'easter. There was seasonably cold air in place over the Northeast and Mid-Atlantic. There was also a ridge of high pressure over southern Canada, extending into the North Atlantic. The high-pressure center was showing signs of weakening.

By Friday morning, the NAM and HRRR (high-resolution) models, which forecast over a short period (several days), had diverged. The HRRR took the center of low pressure northward, hugging the New Jersey coastline. This solution would result in a short period of accumulating snow, followed by a quick change to rain and southeasterly winds bringing in warmer ocean air. The NAM took the nor'easter right along the 40–70 benchmark, the birthplace of East Coast snowstorms. The global models had diverged also; the ECMWF (European) model saw a colder, snowier solution, like the NAM. The GFS and UKMO had the low center about fifty miles offshore, which would mean accumulating snow, perhaps three to five inches accumulating along the coast, with eight to twelve inches accumulating over the colder interior sections.

The nor'easter was forecast to impact New Jersey and the Tri-State area Sunday. By Saturday afternoon, the models were still not in agreement, and time was running out. The National Weather Service had issued a winter storm watch for New Jersey for Sunday, and a winter weather advisory was issued for coastal sections. You are the meteorologist. This is a tough one. What would your forecast have been?

Winter storms are the most difficult of all forecast challenges. Will the cold air remain in place? Will the storm system track inland, bringing rain? Will a short-wave separate system merge with and energize the coastal low and result in explosive deepening? Where will the rain-snow line set up? Often, a mere ten miles separates puddles from snowblowers. How often has Newark reported 3.4 inches of snow while Chatham, not that far inland, has measured 10 inches?

Meteorologists are there to provide—often under intense pressure—the most reliable information possible to the millions of people whose lives, families and businesses would be disrupted by a major winter storm. We owe them a debt of gratitude, not scorn!

This book takes the reader a step back in time as we visit sixteen historic winter storms that left a memorable and profound impact on New Jersey. It contains narratives from the most recognizable meteorologists in the New Jersey and New York metropolitan area, such as meteorologists Joe Cioffi, Craig Allen, Jim Witt and Dan Zarrow, along with the storm stories of electricians, businessmen, realtors, housewives and truck drivers.

Enjoy the adventure!

1

THE BLIZZARD OF '88

In the age of social media and nonstop television and radio storm coverage, with endless forecast revisions and hour-by-hour forecast data being furnished by computer models we've come to know by name, the approach of a significant winter storm is heralded and brandished to a public oversaturated with information. All of this occurs often a week before the first snowflakes descend.

Winter storm watches, or blizzard watches, are issued by the National Weather Service and then upgraded to warnings if the storm threat becomes imminent. Towns and municipalities and large metropolitan areas prepare extensively, with tons of salt and hundreds of snowplows and salt spreaders ready to descend on streets and highways, trying to stay one step ahead of the storm and keep travel safe.

Blizzard conditions on a New York City Street during the blizzard of March 12–14, 1888. Snowfall fell at a rate of two to three inches per hour. *Library of Congress.*

New Jersey and neighboring states often issue a statewide "state of emergency," prohibiting travel entirely with exceptions for emergency situations. Tens of millions of citizens are fully informed about the pending storm hazards, with ample time to shop for food and make necessary storm preparations.

It wasn't always that way!

In the early 1800s, farmers relied on their powers of observation, which were often excellent forecast indicators. Changing leaf colors and the nesting and food gathering habits of squirrels and groundhogs signaled the arrival of autumn. Changing wind directions and the arrival of lowering and thickening clouds usually heralded rain or snow. The wooly bear caterpillar, called the wooly worm in the southern United States, foretold the severity of the upcoming winter with the length of its black bands—the longer the bands, the harsher the winter will be. If the middle brown band is wider, expect a mild winter with less snowfall. Occasionally, a traveler or mail coach would bring news of a snowstorm in nearby states, yet reliable information was nearly impossible to obtain. You were on your own!

In 1844, Samuel F.B. Morse sent the first message by telegraph from Washington, D.C., to Baltimore: "What hath God wrought." This came after a $30,000 appropriation from Congress was narrowly approved to support the invention. A new means of communication was born. The actual invention of the telegraph occurred a decade earlier, with Sir William Fothergill Cooke and Sir Charles Wheatstone introducing the first means of telegraphy, the Cooke-Wheatstone electrical telegraph.

During the Civil War period, the telegraph became a means of transmitting weather information from one area to another as the science of meteorology evolved. Joseph Henry, the first director of the Smithsonian Institute, could actually be characterized as the nation's first meteorologist. He foresaw the need for regional collaboration in sharing weather observations and data. Working with the United States Navy, he and meteorologist Joseph Espy organized a group of observers who shared and communicated observations and information. Weather information was sent to the Smithsonian using telegraphs. A rudimentary means of weather forecasting had been established.

The first official United States government–recognized national forecasting entity was the Division of Telegrams and Reports for the Benefit of Commerce, or Signal Service, under the jurisdiction of the Department of War, signed into law by President Ulysses S. Grant on February 9, 1870.

Although a governmental forecasting agency was a major advancement in promoting public awareness, weather forecasting remained in its infancy. There was no means whatsoever of anticipating storm development, recognizing dramatic temperature shifts or identifying rapidly intensifying cyclones. So, on a mild Saturday morning, March 11, 1888, with people walking about and enjoying a preview of spring with early flowers blooming, there was absolutely no hint that the most powerful and deadly winter storm of all time was less than thirty-six hours away.

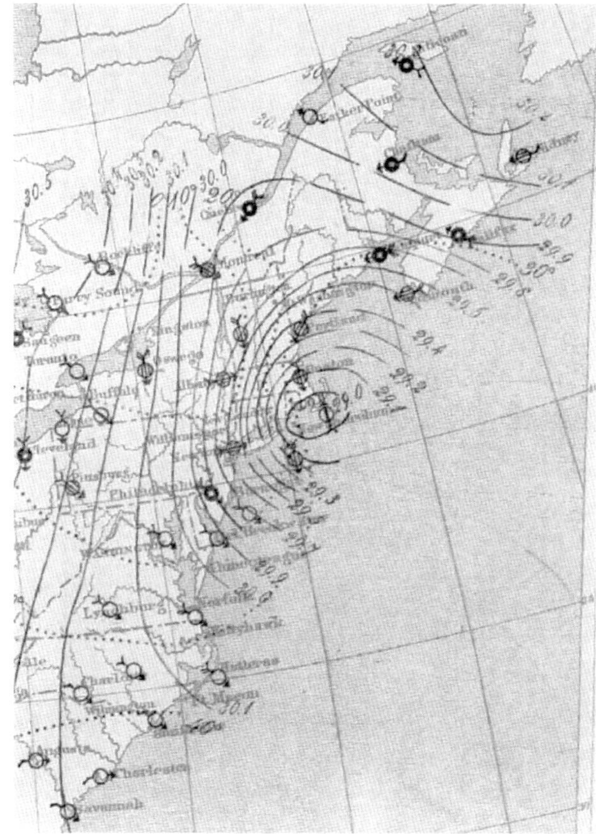

Synoptic chart, March 12, 1888, U.S Eastern Seaboard. *NOAA Library.*

New Jerseyans—and everyone from Maryland to Maine—were truly at winter's mercy!

An elongated frontal boundary, extending southward from an area of low pressure over the upper Midwest, had ushered in a mild air mass over New Jersey. On Saturday, March 10, afternoon temperatures were in the mid-fifties. Newspaper forecasts, usually consisting of a short sentence or two, called for mild temperatures and a chance of rain.

Sunday, March 11, dawned cloudy and mild. Light rain covered New Jersey by the afternoon. By then, the weather was mild, with temperatures remaining in the lower fifties. Though cold air was not present over northern New England, there was a large arctic air mass over central Canada, a ready source of bitter cold air. In the modern world of computer model ensembles and forecast technology, the development of this deadly storm would have been foretold days in advance, much like the 1993 "Storm of the Century." By Sunday evening, an area of low pressure had begun intensifying dramatically

Children help move a mountain of snow near Newton Courthouse after the worst snowstorm since the blizzard of '88. *Courtesy of Wayne T. McCable, president of the Sussex County Historical Society and National Park Service–qualified architectural historian.*

off the Virginia coastline. The "Great White Hurricane" was born. The storm underwent bombogenesis, or rapid intensification, with minimum low pressure lowering to 980 millibars, driving very cold air southward into New Jersey. Rain fell through Sunday night as the freshening winds turned northeast, and arctic air plunged into the region. By midnight, in north-central and Northern New Jersey and New York City, the rain changed to a driving sleet and then to heavy snow after midnight. The thermometer fell into the mid-twenties by 3:00 a.m. Monday, with sustained forty- to fifty-mile-per-hour winds hammering the snow onto carriages, buildings and telegraph poles.

With unbridled ferocity, the intense storm assaulted New Jersey. In central and Northern New Jersey into New York City and southern New England, from early Monday through early Tuesday afternoon, blinding snow and wind covered rooftops with thirty- to forty-foot drifts. Sustained winds, fifty miles per hour, with hurricane-force gusts, resulted in a communications breakdown from Washington, D.C., northward. Rail and elevated lines were immobilized. The effects on mass transportation from this storm were transformational, with an underground subway system proposed in New York City that became operational at the turn of the century.

Date/Time	Temperatures (°F)	Wind	Pressure	Conditions
March 11 12 PM	42	E-SE 12	29.98" F	cloudy
March 11 6 PM	38	E-NE 13	29.81" F	rain
March 12 12 AM	33	NE 25	29.73" F	rain
March 12 6 AM	24	W-NW 50	29.62" F	snow
March 12 12 PM	14	NW 33	29.48" F	snow
March 12 6 PM	12	NW 45	29.50" S	snow
March 13 12 AM	8	N 50	29.40" F	snow
March 13 6 AM	6	W-NW 40	29.24" S	cloudy
March 13 12 PM	9	NW 25	29.28" R	partly cloudy
March 14 12 AM	12	NW 20	29.42" R	clear

Unofficially, four hundred people perished, with two hundred causalities recorded in New York City alone. Prominent United States Senator Roscoe Conkling, the last U.S. senator confirmed by the United States Senate for appointment to the Supreme Court (he declined the nomination), refused what he considered an exorbitant cab fare. He chose to challenge the blizzard and walk uptown. He collapsed from exposure and died from the effects of the storm on April 18.

Two hundred ships were grounded and over one hundred seamen died in the face of the blizzard. It was, beyond doubt, the most extreme East Coast winter storm in recorded history.

Christopher C. Burt

Christopher C. Burt is a weather historian and the author of *Extreme Weather, a Guide and Record Book*, a work about extraordinary weather events and how they altered history. He studied meteorology at the University of Wisconsin–Madison. He was a contributing writer to the *New York Times* and *the Los Angeles Times* and is currently employed by the Weather Company in San Francisco, California.

> *As Paul Kocin and Louis Uccellini noted in their classic compendium* Northeast Snowstorms, *the Blizzard of '88 was unique for several reasons. Firstly, most severe winter storms that affect the Northeast are preceded by an outbreak of cold air across the eastern United States, usually centered over northern New England or southern Canada. No such air mass was in place prior to the development of the storm. Secondly, the storm center became stationary and actually made a counterclockwise loop off the coast of southern New England while maintaining its peak intensity (with a central pressure of approximately 980* [millibars]*). Instead of moving along the usual southwest-to-northwest path that severe winter storms tend to follow, the low-pressure center just gradually filled in and dissipated, eventually drifting slowly out to sea. This map illustrates the track of the Blizzard of 1888 and how it meandered for forty-eight hours off the southern New England coast.*
>
> *In New York City, the rain turned to snow at 1:00 a.m. on Monday, March 12, when the temperature fell to freezing. Blizzard conditions quickly developed as the wind rose to a sustained 50 miles per hour. By 8:00 a.m. Monday, the city was completely immobilized by the blinding, drifting snow and howling winds. All telegraph communications went down. There was no subway at the time, and the elevated rail line ground to a halt, with one train derailing and killing several passengers and crew.*
>
> *Walking in the streets became not only impossible but also deadly. Of the two hundred people who perished in New York City, most were found buried in snow drifts along the city's sidewalks. One of these victims was Senator Roscoe Conkling, a New York Republican Party kingpin and aspirant for*

The blizzard of March 12–14, 1893. As twenty inches of snow cover a Manhattan Street, three children make their way to the iron foundry, industry typical of the late nineteenth century. *Courtesy of NOAA.gov.*

A horse-drawn sleigh navigates city streets in steady snowfall during the blizzard of March 12–14, 1888. *Courtesy of NOAA.gov.*

the U.S. presidency. He died as a result of "over exposure" from trying to walk from his Wall Street office to the New York Club on Madison Square. Refugees filled all the hotels. The venerable Astor Hotel set up one hundred cots in its lobby when it became apparent by sunset that day that venturing outside was still impossible. The temperature had fallen to [eight degrees Fahrenheit] *by sunset; the wind was still howling, and snow drifts up to twenty feet filled the streets of the city.*

The storm was even more severe in areas north and east of New York City. Fifty trains became stranded between Albany and the city, as well as on Long Island, in New Jersey and in Connecticut. Many were derailed after trying to plow through drifts measured up to thirty-eight feet in Connecticut (this drift measured in a rail line cut near Cheshire). Drifts up to forty feet were reported in Bangall, a small town in Dutchess County, New York. Many of the two hundred fatalities attributed to the blizzard outside of New York City consisted of passengers and train crews that attempted to walk to nearby towns after their trains became stalled or derailed. Several ships foundered at sea, lost to ninety-mile-per-hour winds, huge seas, and ice accumulations on deck that caused them to roll over from the top-heavy weight.

METEOROLOGIST DAN VALLE

Dan Valle is a lead forecaster for the National Weather Service. He provides forecast information for many regions in the United States, including the mountains and Snake River Plains region in Idaho. He is a weather historian, having written the article, "The Blizzards of 1888" for National Weather Service Heritage.

The weekend of March 10, 1888, started off rather pleasantly in the Northeast: Saturday brought early spring weather, complete with growing grass, chirping birds and budding trees. However, by Sunday afternoon, the temperature had suddenly dropped, and rain began to fall. Come Monday morning, the rain changed to snow and the warm breezes transformed into powerful gusts of at least fifty miles per hour. Before long, the snowfall amounts reached forty to fifty inches, with snow drifts between thirty and forty feet deep.

The storm cut off and immobilized East Coast cities, crippling transportation and affecting one quarter of the U.S. population. The storm became legendary in New York City: as the economy was struggling, most workers went to their jobs regardless of the weather conditions. As a result, there were numerous accounts of people stranded and freezing to death. More than four hundred people died from this storm, two hundred in New York City alone. Additionally, the winds were so fierce that more than two hundred vessels were destroyed up and down the Eastern Seaboard, resulting in the deaths of one hundred seamen.

The failure of the Signal Service to issue a "Cold Wave Warning" for these two calamitous blizzards became a motivating factor for moving the meteorological service out of the War Department so as to improve forecasting and preparedness efforts. Two years later, the legislation creating the Weather Bureau under the Department of Agriculture was signed by President Benjamin Harrison on October 1, 1890.

2
THE BOXING DAY SNOWSTORM OF 1947

By 1947, forecasting accuracy had improved significantly. However, the science of meteorology and weather forecasting was still antiquated when compared to today's technologies. There was no way to identify the phasing of the northern and southern jet streams, which usually result in major storm development. There was no enhanced radar and no means of identifying an unexpected shift in a storm's trajectory.

It was the postwar era, yet citizens remained at winter's mercy!

Forecast updates occurred usually twice daily, and since television was still in its infancy, people got their weather forecasts from the radio and morning newspaper. During World War II, printed newspaper forecasts were either eliminated entirely or reduced to only a word or two, for fear the Germans would use the information to plan attacks on the U.S. mainland. There were no hourly storm updates, no television coverage, no social media. If a snowstorm seemed imminent for New Jersey, a heavy snow watch and then a warning would be issued. But this did not occur on December 25, 1947. The National Weather Bureau had no clue what was unfolding.

Moderately cold air was in place over New Jersey, a storm system that had developed over the Gulf states and moved offshore, parallel the Eastern Seaboard, supported by a large plume of Gulf moisture. No explosive storm developed. The warm moist Gulf air collided with the cold air in place, and a large area of heavy snow blossomed over New Jersey and New York City into Long Island. What was unique about this snowstorm without wind was that the heavy snow retrograded into New York City and New Jersey, moving east to west off the Atlantic coast instead of in the usual south-to-north movement.

Snow developed at 3:00 a.m. Friday morning, December 26, and quickly became heavy. Snow fell relentlessly at a rate of one inch per hour all day and through the night. The weather bureau could not keep pace with the storm, increasing accumulation amounts again and again. In terms of ferocity, the storm did not rival the Blizzard of '88, yet it overwhelmed cities like Newark, Elizabeth and Paterson, which were caught utterly off guard the day after Christmas.

Northern New Jersey and New York City were squarely in the bullseye. Newark received 26.2 inches of snow, Long Branch received 29.7 inches and Central Park in New York City recorded 26.4 inches. Farther south, snow accumulations were smaller, with Trenton receiving 8.2 inches.

The surprise massive snowstorm claimed thirty-one lives in New Jersey. Bus and rail lines ground to a standstill. People were stranded, forced to take refuge in bus depots or churches or at work. It was a stark reminder of how weather forecasting, in 1947, remained clearly in its infancy. Not one forecast had a clue that a historic snowstorm was bearing down on New Jersey.

CHARLES GAFFNEY

Ken Gaffney was an employee of Public Service Electric and Gas, working the poles, servicing Central New Jersey. He was hired as a utility worker shortly after the war ended, having served in the United States Navy in the Pacific theater. Ken, like thousands of patriotic young men, enlisted in the Navy before graduating high school, against his parents wishes.

His father, Charlie, was a long-time postman for the United States Postal Service. His postal route extended through Roselle, Cranford into Westfield, an arduous fourteen-mile trek he made until his retirement at age seventy in 1957.

Charlie, his wife, Mae, and Ken spent a quiet Christmas Day on Thursday, enjoying the traditional Gaffney dinner of roast duck, coffee and cake. Charlie's fondness for dessert earned him the nickname, given by Mae, "Coffee and Cake Gaffney." It never showed; he weighed 154 pounds. The day grew overcast, yet there was no concern about the weather.

By Friday morning, five inches of snow was on the ground, and it was snowing heavily, vertically, not a lick of wind. Ken had to report for work on Friday, navigating his way through snow-clogged, nearly impassable roads.

The snow had taken wires down, hundreds of county residents were without power and Ken's crew went from one job to the next. There were no snow days in 1947, unless you wanted to forfeit a day's wages.

Charlie had taken the week off as vacation time; the Gaffneys had planned a weekend holiday trip to Point Pleasant to visit Tante (Aunt) Lena. The snowstorm caused the visit to be postponed, even though Point Pleasant had received far less snowfall. Charlie couldn't even get out of Linden, let alone navigate the pre–Garden State Parkway roads going down the shore. His 1937 Plymouth was unrecognizable, buried under a mountain of snow. It took all of Friday, day and night, to clear the sidewalks and driveway and shovel two feet of heavy, wet snow from the Plymouth.

Meteorologist Jim Witt

Jim Witt is a long-time nationally recognized meteorologist, educator and writer, and he has been an inspiration for aspiring meteorologists for over sixty years. Born in Queens, New York City, Witt has a passion for weather that was born when a hurricane swept up the East Coast in 1944, when he was seven years old. It was a passion that lasted a lifetime and has inspired thousands of people, both young and old, to pursue careers in meteorology.

Jim works every day, developing short- and long-range forecasts for his acclaimed "Weather Wiz" program. He is, as of this writing, eighty-seven years old, recognized worldwide as a trailblazer in the field of meteorology.

From the Lakeland School District in Westchester County, New York, where he developed and instructed Advanced Meteorology, to creating and operating a now nationally acknowledged secondary school weather program, the "Weather Wiz," Witt has mentored and inspired the careers of many recognized meteorologists and personalities in forecast broadcasting.

In 1967, Witt joined the faculty at Columbia University, where he organized, taught and directed meteorological programs for both Columbia and the National Aeronautics and Space Administration (NASA). Witt was also an instructor of meteorology at Western Connecticut State College and Pace University. A pioneer in the science of long-range forecasting, Witt provided WCBS-TV, WABC radio and WOR radio with long-term forecasts. In 1982, he discussed long-range forecasting as a guest on the long-running MacNeil-Lehrer report.

Witt was also a broadcast meteorologist on WHUD 100.7 FM in Beacon, New York. His most enduring legacy was producing the Hope for Youth Foundation Long-Range Weather Calendar, which is currently in its fortieth edition; $6.8 million in proceeds from the calendars have been given to sick and terminally ill children, and it provides eighteen college scholarships yearly to deserving students.

The "Weather Wiz" program was started by Witt in 1962 to develop and inspire young people to take an interest in meteorology. Through a weather book, step by step, youngsters learn how to forecast weather. They discover that weather is fun and that forecasting is an exciting career.

> *I became a weather nut in September 1944; it was September 15 when a great hurricane swept up the East Coast, bringing torrential rain and wind to New York City, Long Island and New Jersey. Trees were down everywhere, houses were smashed. I was absolutely glued to the barometer; it was so exciting.*
>
> *I was ten years old, Christmas 1947. I lived in Middle Village in Queens. It was a good neighborhood with big, open lots where we played baseball and football. It didn't look like we would have a white Christmas. I followed the daily forecasts on WNEW radio, and on Christmas Eve, they called for flurries on Friday, the day after Christmas. On Christmas Day, the papers and radio still said there would be no snow.*
>
> *I also listened to the* Rambling with Gambling *radio program on WOR to get the daily forecasts. In fact, years later, I was a guest on the program every first of the month to give listeners my long-range forecast. Back then, the program got their daily forecasts from the Weather Bureau. On Christmas Eve, flurries were mentioned—that's all. John B. Gambling hosted the program, which continued for ninety years, with three generations of John Gambling's at the microphone. WOR was a part of the old Bamberger's department store in Newark.*
>
> *I went to bed Christmas night, and when I woke up, there was five inches of snow on the ground, and it was snowing heavily. I never saw such terrible forecasting in my life. The day before, they said there would be no snow. The Weather Bureau could not keep up with the heavy snow. First, they forecast six inches; there was already eight inches on the ground. By afternoon, they said ten inches; there was over a foot on the ground.*
>
> *The snow never let up, not once. I spent all day Friday peering out the window as the snow piled up. The city was shut down, cars covered to the rooftops, people stranded. Buses were stuck in the snow, barely moving. It*

was a mess. New Jersey was hit hard also; I had an uncle in Califon who told us there was nearly two feet.

My dad had to walk home through the blinding snow from Manhattan to Queens. He left work at 4:00 p.m., and it took him five hours to get home. The snow was up to his knees. There were no cellphones back then, and everybody was worried. We ended up with twenty-six inches of snow in twenty-four hours. Continual, relentless, heavy snow. When it got dark, I watched the huge, wet snowflakes in the shadows of the streetlights. The snow just piled up; there was no wind, no drifting.

Late Friday night, with snow still falling. I saw the moon visible through the clouds, a sign the storm was almost over.

This storm furthered my interest in meteorology and long-range forecasting, in particular. I have downloaded all the past weather for about 780 cities in the United States and many more around the world. Using this data and a mathematical formula for determining the exact location of the sun, moon and planets to the minute, out one thousand years and back one thousand years. Go to www.theweatherwiz.com and input a date at least one week in the future; it can be three years from now, and you'll get the forecast for that date.

Wayne Blanchard

Wayne Blanchard has created a webpage, Deadliest American Disasters and Large Loss-of-Life Events, an educational site citing news articles and commentary on deadly American disasters and other large loss-of-life events. It's a factual and statistical resource for disasters, fatalities, injuries and destruction in the United States.

New York (AP): The storm, sweeping in with surprise fury early yesterday, surpassed lot 20.9-inch downfall left by the famous Blizzard of 1888. It pelted the area with an average 1 hourly fall of 1.8 inches and ended officially after 15 hours and 45 minutes. At least 35 persons lost their lives in the storm belt which embraced parts of New England, Pennsylvania, New Jersey and southeastern New York and extended south to include Washington, D.C. New Jersey, where 30 inches of snow was officially reported in Long Branch, counted 12 dead, and New York 9. Connecticut reported 5 deaths, Pennsylvania, 2; Massachusetts, 3; Rhode Island, 2; and New Hampshire, 2.

3
THE GREAT ATLANTIC STORM OF 1962

Helicopters rescued two hundred people stranded in Cape May. Avalon and Sea Isle City were devastated. Damages to the beachfront at Atlantic City were $5 million, surpassing the 1944 hurricane. Storm damages at Ocean City reached $10 million. Eleven men were reported missing from two New Jersey fishing trawlers.

Weather forecasting had advanced a great deal by the early 1960s. With the launch of the first TIROS-1 television infrared observation satellite on April 1, 1960, followed by the launch of the first NIMBUS weather satellite on August 28, 1964, forecasters were provided with imagery and atmospheric data that served to further improve and refine forecasting. On the weekend of March 3–4, 1962, satellite imagery and available forecast data provided barely a hint that the most catastrophic nor'easter in recorded history was mere days away from ambushing the Mid-Atlantic and Northeast.

There aren't adjectives to describe the destruction that was visited upon the entire New Jersey shoreline from Tuesday to Thursday, March 6–8. A low-pressure area was bringing accumulating snow to portions of the Ohio valley. A ridge of high pressure was building over the far Northeast and Atlantic Canada, extending over the shipping lanes and providing a block for any storm system moving northeastward. Low pressure along an associated frontal boundary developed along the Southeast coast. The upper-level system, moving southeastward, merged with the intensifying nor'easter. There were no GFS or European models, no 120-hour short-range forecasts to put all the pieces of the puzzle together. No warnings!

The Weather Bureau forecast for Monday, March 5, in Newark was increasing cloudiness, a chance of some rain Tuesday, followed by clearing and seasonably cold temperatures. The great nor'easter deepened into a cyclone comparable to a category 1 hurricane. Cold air draining into the circulation of the massive storm brought heavy snow as far south as Alabama and into interior North Carolina and Virginia. The Outer Banks of North Carolina were mercilessly battered, with thirty-foot waves crashing against the fragile barrier islands, cutting a new inlet between Avalon and Buxton.

The sprawling storm, which would be classified as "level 5" on the Dolan Davis scale of extreme nor'easters, began crawling to the north and east, its forward motion blocked by the large ridge of high pressure extending over the North Atlantic. Heavy snow and rain overspread the New Jersey coastal plain on Monday. A six-hundred-mile fetch of northeast winds, between the high-pressure ridge in the North Atlantic and the powerful nor'easter, sent walls of water onto the New Jersey coastline, devastating beach communities. Catastrophic destruction was experienced from Barnegat Inlet southward, obliterating thousands of beach homes, some ravaged by fire. An estimated forty-five thousand homes were destroyed or damaged, a vast majority along the shoreline; 80 percent of the homes on Long Beach Island were lost. There were five high tide cycles, each one smashing beach homes and sweeping them into the ravaging ocean.

Harvey Cedars, Long Beach Township, Beach Haven, Ship Bottom, Avalon, Sea Isle City, Loveladies and Holgate, all seaside towns on Long Beach Island, were wrecked by the crawling nor'easter. Almost every resident of Sea Isle City was told to evacuate as floodwaters five feet deep inundated the town, destroying landmarks like the Windsor Hotel, Madeline Theatre and the Excursion House. Inlets, where none existed previously, were the byproducts of the storm on Long Beach Island.

Statewide storm damages were estimated to cost $80 million—later estimates put the figure near $100 million. Storm damages at the shoreline and barrier islands totaled $30 million. There was a human toll: thirty-two New Jerseyans died in the storm's path. Sustained tropical storm–force winds of fifty to sixty miles per hour lashed the New Jersey coastline, with a seventy-three-mile-per-hour gust reported at Long Branch. Portions of the island were entirely underwater. Houses were bludgeoned off their foundations, and dozens of families were trapped and subsequently rescued, many by local police and fire services.

Farther north, on Barnegat Beach Island, communities from Seaside Park north to Point Pleasant received significant yet less destructive

Beaufort's Scale (for measuring wind)

Force 0, calm (0 MPH)
Force 1, light air (1–3 MPH)
Force 2, light breeze (4–7 MPH)
Force 3, gentle breeze (8–12 MPH)
Force 4, moderate breeze (13–18 MPH)
Force 5, fresh breeze (19–24 MPH)
Force 6, strong breeze (25–31 MPH)
Force 7, moderate gale (32–38 MPH)
Force 8, fresh gale (39–46 MPH)
Force 9, strong gale (47–54 MPH)
Force 10, whole gale (55–63 MPH)

impacts than Long Beach Island. Three thousand feet of the boardwalk at Longport were destroyed; one thousand feet of the walkway at Long Branch were lost to the ocean. The precipitation began as a mixture of rain and snow over Southern New Jersey. It fully snowed for a time from Long Beach Island northward before changing to rain. New York City had light snow most of the day on Monday, and Philadelphia International Airport recorded 6.8 inches of snow.

Margaret Thomas Buchholz

Great Storms of the Jersey Shore, coauthored by Margaret Thomas Buchholz and Larry Savadove, is the definitive narrative of the Great Atlantic Storm of 1962. The book was first published in 1993, with a second edition released in 2019, which also was coauthored by Scott Mazzella. It was a 2020 Gold Award Winner, named the best regional book in the nation.

Margaret Thomas Buchholz is the voice of the history of Long Beach Island, having authored *New Jersey Shipwrecks: 350 Years in the Graveyard of the Atlantic* (2004), which won the Foundation for Coast Guard History award ("A brilliantly researched chronicle of shipwrecks."); *Island Album: Photographs and Memories of Long Beach Island* (2006); and *Josephine: From Washington Working Girl to Fisherman's Wife* (2012), a luminous biography of a spirited woman and her journey through the first half of the twentieth century.

After she was born in Manhattan in 1933, her parents relocated to Harvey Cedars on the northern end of Long Beach Island during the Great Depression. Margaret lived much of her life on the barrier island and was the owner and publisher of *The Beachcomber*, a free island newspaper that became a part of the daily life of island residents from 1955 to 1987. Her nickname, "Pooch," was given to her as an infant by her father; it was a sobriquet that followed her throughout her life.

> *There is no story to this storm. It didn't have what's needed for a story—a beginning, middle and end. It was all middle. It came all at once, without warning. It stayed for three days. Then it went away. It was nothing like a hurricane, with notices and watches and advisories to herald the approach, flags flying, satellite pictures of the beady little eye and flailing arms.*
>
> *Is it heading here? Is it heading out to sea? And then it comes and blows and bellows and spits and rages, and then the calm eye passes over—God's eye, some have called it—and then you get hit with the backside, furious, mighty, unassailable, unanswerable, and then it goes off to unload on someone else.*
>
> *A hurricane is a progression: the sound of distant cannon, the attack, a lull, redoubled violence and then peace. This storm was nothing like that. The storm was five high tides, the highest 8.6 feet, just 4.8 inches below the 1944 9-foot record, water that just kept coming. It was a nor'easter more ruinous than any hurricane that ever happened here, a demonstration that nature is a hard mother, a reminder that those who dwell beside the sea are always only a wavelength away from sleeping with the fishes.*
>
> *"A savage gale with pounding tides, towering seas and heavy snows," said the* New York Times.
>
> *It was unexpected and unannounced, sudden and surly, inundating, devastating, mutilating, obliterating. It battered and bludgeoned the shore until there was no more shore, until it was all running water and milling debris, until almost every trace of human presence had been washed away. Then it was gone.*
>
> *But not, to this day, forgotten. One newspaper called it "a scene of human misery unparalleled in the memory of longtime resort residents," exceeding that wreaked by the hurricane of 1944.*
>
> *Correspondent Robert Holland was one of the first to fly over the area. "You can see giant puddles and gashes where the ocean laps at the bay in places that used to be part of Long Beach Island. A north section of the*

island below Harvey Cedars reminds you of what would happen if a little boy swept his hand across a sand pile, crushing his toy roads and trucks."

Back on the ground, he began collecting stories of the storm, citing the heroic and the not-so.

Such as an excitable Philadelphian who somehow caught the second helicopter out, ahead of most of the women. This gentleman said he would pay "whatever it costs" to get to the mainland and offered to buy, from weary women who had been handing out free food for two days, meals for his rescuers. The Marines from Long Beach School had a name for him.

I hitched a ride onto the island on the back of a high wheeled utility truck. We passed a housing development east of Mud City that proclaimed, "Every Home on Large Waterfront Sites."

Charles Roberts, seventy-seven, took a last look as he headed the other way. "Everything's gone," he whispered, "everything's gone." But it was over. Under the headline, "Nature Lifts Angry Hand from Storm Torn Shore," reporter Victor Keller wrote.

"The mood of the island people was almost festive. The terror ended in a kind of reckless gaiety, like a peace following a war. People who had stayed on the island mostly gathered together to eat and talk, counting the big blessings, discounting the loss. Thousands of others evacuated to the mainland camped out in school buildings and made a game of it. The reckonings had not yet begun."

Flying over Sea Isle City, Adolph Wilsey said, "This resort looks like a huge lumber yard." Reporting from Wildwood, Steve O'Keeffe said, "I never saw such a display of sustained power and fury." Far across the Atlantic, the coasts of Ireland, Southern England and France were battered by gales that churned up waves as high as fifty feet for two days. On this side, some four hundred bulldozers went to work to create a new dune line as people began sifting through the sands.

Marion Rapp found a Bixby ink bottle sitting on a stump in front of her house. It bore the inscription, "March 6, 1883." For almost eighty years it had lain under the sand waiting to be uncovered. The rudder from an old sailing vessel, twelve feet long, eight inches thick and clad in its original copper, washed up at Holgate, not far from where the Monssen itself lay beached.

Dave Wood found George Otto's Coke machine in the bay the next summer in about six feet of water, "still full of Cokes, plus six bucks in quarters. Those old, thick, green bottles—you can't kill 'em." Wood and his friends spent the summer poking around in boats.

"Cars. I remember just south of Gunner's point, this big black car, sunk there. It was very ominous. To this day, when I get near something that's sunken and manmade like that, a sunken boat or wrecked thing, it gives me the heebie-jeebies. We were kids, having a good time. There were a lot of wrecked houses lying around. You could just go play in them. And in the bay, too. We rowed through one night, right through it, it was a lot of fun.

"I got cut constantly that summer, walking in the bay and stepping on stuff. But everything was a treasure. I still have one weird thing, sort of a black, lacquered affair, some kind of Japanese art thing, I suppose, very exotic.

"We watched a gang of kids pour gasoline in one half sunk house. They spent all day chopping holes in the roof to ventilate it so it would really go up on July 4th. It was great."

The U.S. Coast and Geodetic Survey sent a team to study Little Egg Harbor, Absecon and Barnegat inlets and started recharting the shore immediately. Captain Lawrence Swanson said it was the first time a major survey has been prompted by a single storm, calling it "one of the biggest natural changes this century."

By the end of July, the New York Times *reported:*

> There is still much evidence of devastation, but the comeback has been spectacular. Row upon row of neat, brightly painted cottages, motels, apartment units, stores and concessions, have been erected to replace structures washed out to sea.

There were flood damage sales, but property values did not drop, as some had predicted. People kept finding things all summer long. Bottles of liquor turned up, seals still intact but labels gone. "The only trouble was," remarked an islander, "if you found something you liked, you didn't know what it was."

Stephanie Schantz found a gold ring in North Beach, "Up by the Coxes, what was left of their house. I tried to get in touch with them, but could never find them. I still got it. I keep it in my jewelry box." Everybody kept a memento, some leaving a wall unpainted to show visitors the high-water mark. The most popular souvenir seemed to be a set of 105 storm photos that Bill Kane of the nor'easter shop in Beach Haven Terrace put together and was selling for fifteen dollars.

The Great Atlantic Storm left its stories scattered amidst the debris. And its memories. Betty Hornby wonders, "How many remember it as I do." Chief (Angelo) Leonetti and his wife were so much fun. He called her "the bride," even though they'd been married for thirty years. Bob Osborne was kind to all kids who hung around his magazine racks, reading everything they could and never buying; he never seemed to mind. Kenny Chipman, always active, always helping those in need. They have been washed away with the sands of time.

The Beach Haven Times *put the sentiment of a lot of people into an editorial:*

> Tuesday, February 20, 1962, the culmination of man's mastery of science sent a man out of the earth and ionosphere to circle the world. Tuesday, March 6, 1962, man's puny scientific skill in predicting the vagaries of nature brought incredible terror, destruction and death to a strip of this earth. If scientific calculations can send us into orbit, why can't they warn us as to what's going on down here.

SCOTT MAZELLA

Scott Mazella, author of *Surviving Sandy: Long Beach Island, and Great Storms of the Jersey Shore*, is a history teacher, reporter and alumnus of Rutgers School of Communication and Information (SC&I), where he majored in journalism and media studies. Scott has had an interest in meteorology since his youth. His parents had a shore home in Holgate, a South Jersey beach town that was extensively damaged during Superstorm Sandy. Comparisons between Sandy and the Ash Wednesday storm have grown, taking into consideration the substantial population growth on the barrier islands since 1962.

If we had a three-day nor'easter today, like in '62, it would no doubt create unprecedented damage and destruction, especially in vulnerable areas along the coast. It is possible there would be new inlet-like cuts through towns as we saw in '62 and also during Sandy. Infrastructure would be impacted, roadways and highways impassable in places.

Dune replenishment, what is still intact, will buy towns precious time but, if breached, will open the door to disaster. A '62 storm, if it occurred today, would create damage on par with Sandy, perhaps even greater, depending on the strength of the storm and how many homes and businesses survive. Hopefully, building improvements and the raising of homes since Sandy will help limit the overall damage.

PAT HAFFERT

Pat Haffert, seventy-five, is a native of Sea Isle City. He is a storm survivor of the Great Atlantic Storm of 1962 and a member of the Sea Isle City Historical Society. He was born in Mercy Hospital on the barrier island in 1949 and was ten days short of his thirteenth birthday when the powerful nor'easter, coinciding with the annual "spring tide," unleashed its fury, with five deadly high tide cycles bringing catastrophic destruction to the barrier island.

Haffert's family predates the Great Depression on the island. His grandfather, a former three-term mayor of Sea Isle City, was a newspaperman who later formed the Garden State Publishing Company, which produced trade magazines for the poultry business. During his tenure as mayor, Haffert would not support a proposal to rezone Railroad Avenue for residential construction in order to raise revenue and increase tax ratables.

For years, the dead-end boulevard was derisively referred to as "Haffert's Folly." However, the courageous stand Haffert took generations ago is reflected today in John F. Kennedy Boulevard, a scenic, broad entranceway with ocean views from the top of the bridge. It is one of the most beautiful gateways onto any barrier island in New Jersey.

Haffert and his wife, who worked for the FDIC, lived in Washington, D.C., for several years, and he traveled throughout the East Coast selling advertising space for poultry magazines published by his family-owned publishing company.

He later transitioned to the advertising business, first with the H.T. Fenton Advertising Agency in Somerville, New Jersey, and later with Bozell & Jacobs in New York City. Haffert continued to work for various businesses in his "business-to-business" marketing career, with his last position being at Danbro Distributors in Philadelphia.

> *In the personal recitation of storm survivor stories like mine, consistent, common themes appear: fear and isolation, a lack of information, not knowing what comes next, as the phones and electric were both down, and the difficulty of moving around without a boat. People helped one another; good Samaritans rose to the occasion and offered food, shelter and comfort to the displaced.*
>
> *People moved from one location to another trying to escape the ever-encroaching storm surge. Catastrophic events often bring out the best in people, and that was the case in Sea Isle during the worst storm to ever hit*

the Island. The ocean moderates the temperature on coastal islands, and the salt in the air, particularly if it's blowing from an easterly direction, can combine to lessen the snowfall amounts on barrier islands significantly—if we get any snow at all.

Often, the storms don't get this far east at all, as most snowstorms usually stall somewhere just east or west of the I-95 corridor. So, growing up on a barrier island as kids, we always felt we got cheated out of the major snowstorms blanketing Philadelphia and the suburbs. TV stations were raising the alarm and predicting widespread school closings, but in Sea Isle it was usually just another rainy day at the beach. This time, things were different. My brothers and I awoke to find water around the house and in the driveway.

This was unusual, as the beach side of Sea Isle is higher ground. My mother got a call with some great, unexpected news: both schools were inundated with water and would be closed for the day. We finally got a weather-related break with a day off from school while our summer friends and cousins from Philly had to go to school. When you're a kid, you're in the moment. Today was going to be a good day, and then it even got better. Our home was actually a large apartment carved out of an old hotel/boardinghouse, with two additional apartments and rooms which we rented out in the summer. Our immediate relatives in town had water in their homes and had lost power, so my parents extended an invitation to join us. We had heat, electricity and a front-row seat to watch this developing storm. We were right on the beach at the Surfside Hotel at Fiftieth Street, unexpectedly open for a winter storm watch party! My two cousins, their dog, my aunt and uncle and my grandparents were all coming to the house. Things were only getting better—until they weren't!

We all were gathered together, enjoying one another's company, but you could see that things were getting worse, not progressing in the right direction. In addition to the storm surge, there were other elements to this storm. The incessant howling wind was competing with the roar of the ocean. Sheets of rain were coming sideways, lashing the building and windows. Floating debris smashed into existing structures. Porches, garages and storage sheds were breaking loose and smashing into other structures.

It was very noisy and frighteningly eerie. We had a dog named Runzel, which is German for "wrinkles." He had large, floppy ears and lots of wrinkles. Earlier in the evening, before the tide came up, we let Runzel outside, and he never came back. Now, my father had these whining kids who were crying and worried about the fate of Runzel.

Top: Submerged vehicles at Fortieth and Landis Streets in Sea Isle City, New Jersey, after the third high tide cycle during great Atlantic storm of 1962. *Courtesy of Sea Isle City Historical Society.*

Bottom: Runzel, the survivor dog who safely found his way to a neighbor's house during the great Atlantic storm of March 5–9, 1962. *Courtesy of Sea Isle City Historical Society.*

My father, playing defense, said, "Animals have survival instincts. They are not like people; they know what they are doing; they are really on top of things. That dog is smart; he is headed for higher ground; don't worry about that dog." We were undeterred: looking out the window, we could see there was no higher ground. We had a big picture window, six feet high and maybe eight to ten feet wide, looking right out over the ocean.

The water was hitting the bulkhead, flying straight up into the air; the wind would catch it and blow the spray back into our picture window on the second floor of our apartment. We had French doors down in the lobby of the hotel. The waves broke through, smashing the doors open, and now, the lobby was being flooded by the storm surge. My uncle, my father and my brother Joe, who was seventeen years old, went down to the lobby and tried to barricade the doors to keep the water at bay as several of us kids looked on.

The tide kept rising and the waves were bigger and bigger. There was more and more water on all sides. Our hotel was now an island, completely surrounded by water. You could see that the waves were starting to break through the bulkhead, shattering and splintering it. Without the bulkhead, the surge ran unabated underneath the building.

With every wave, the building shuddered. It was terrifying for all of us, especially for the children. Nowadays, pilings on the beach are jetted in with water to get them deep enough to prevent failure from extreme scouring caused by storm surge. In the early 1900s, when the Surfside (then the Seminole Hotel) was built, foundations for beachfront properties consisted of timber piles, usually the trunk of white cedar trees, merely driven to refusal. That type of foundation is inadequate and not driven deep enough to withstand the almost twenty-foot-high waves which rapidly scoured out and eroded the sand until the support piles were exposed.

This was the phenomena that, unbeknownst to us, was occurring underneath our home, the Surfside Hotel.

It was right around dinner time. The children were going to eat first in the kitchen, and then the adults were going to eat in the dining room. My cousins Mike and Jan, my younger brother, Greg, and I were in the kitchen with my mom and Aunt Jean. While my mother was beginning to serve dinner, the house gave a loud shudder and lurched forward, coming to rest at a fifteen-degree angle. The lights went out. Much like the movie scene from Titanic.

People were thrown to the floor; cabinets and drawers flew open; and glasses, plates, knives and forks exploded out all over the room. My brother

Oceanfront flooding and homes destroyed during the aftermath of the Ash Wednesday storm of March 5–9, 1962, Sea Isle City, New Jersey. *Courtesy of Sea Isle City Historical Society.*

Mark and our grandfather were in the living room watching the action unfold through our picture window. When the house lurched forward, the picture window exploded, shattering into thousands of shards of glass.

Even to a thirteen-year-old kid, the situation was obvious. The building was starting to fall into the sea. At this point, my father revealed our well-conceived evacuation plan. He got to his feet and said to my mom: "Helen, take the children and run." My mother instructed me to take the four youngest children and evacuate immediately to the Bednareks' home, the only other occupied house on the street. "Take the shortcut. Go the way you always go [i.e., through neighbors' backyards]*," she said.*

This was the safest, most direct path. The movement of the house had jammed the door shut, so my cousin Mike needed to break the door to quickly exit the building. My brothers, Mark and Greg, cousin Jan, their dog, "Tubber," and I followed out into the raging storm. Due to the high floodwaters, we took a while to navigate the neighborhood's backyards, fences and floating debris in the pitch dark.

The howling winds and crashing surf were deafening. Terrifying. It was windy, and the waist-high water that we were slogging through was ice cold. Mike and Mark rushed on toward our destination. Jan, Greg, Tubber and I lagged, as Greg was only six years old, and he and the dog needed to be dragged or carried. Mark and Mike, the advance team, arrived first at the home of our neighbors, the Bednarek family, followed shortly by the stragglers.

Cold and frightened, we were anxious to get out of the gale. When we arrived at our destination and opened the door, we were greeted by a familiar face. Runzel had escaped our house and imminent danger and found his way to the Bednarek home. I always joke that, at this stage of my life, my esteem for Runzel, who knew exactly what to do and where to go, goes way up. While, for my dad, who said, "Let's invite the whole family over so we can all die together," goes in the opposite direction!

A different scene was playing out with my parents, grandparents, aunt, uncle and brother Joe. They took more time in vacating the house. However, in the confusion, the men evacuated to my mother's parents' home at Twenty-Sixth and Forty-Ninth Streets, while the women, who exited before them, headed to the nearest refuge, the Bednareks'. My mom and Aunt Jean realized they could not take my grandmother over the fences, so they proceeded through a neighbor's yard but parallel to the storm surge until they reached Fifty-First Street and could turn west toward Pleasure Avenue. Fighting the gale winds and ice-cold, debris-laden, waist-high water, my grandmother decided that she could no longer go forward. Holding onto a signpost, she pleaded with her family to go on without her, as she could no longer make it. She was pulled away from the signpost by her daughters-in-law, and the three women moved on, eventually reaching the Bednareks' home on Fifty-First Street.

The following day, Wednesday, the 7th, the children, sleeping on the third floor, rushed from the front bedroom to the back bedroom and cheered at the sight of the still-standing but notably listing "Surfside," which had weathered the night. My dad showed up shortly thereafter at low tide with a flatbed truck borrowed from the city and evacuated both families before the next high tide. If you chose the proper route (high ground), trucks could navigate at dead low tide; otherwise, it was only [passable] *by boat.*

The next stop for us was the Lewis Funeral Home on Fifty-Third Street. The Bednareks were taken in by the family, who lived next door to the funeral home. Out of the wind and now somewhat dry, we passed the rest of the day and evening. Due to the many broken and exposed pipes, Sea Isle City's water and sewer systems had failed. The funeral home had more bathrooms than the average house, but soon, even their facilities became inoperable and overflowing.

Because of this, it was decided to seek an alternate shelter the next day. Wednesday afternoon, after the high tide, my dad, in consultation with my mom, decided to visit our house to retrieve valuables. Knowing that all the steps leading to the building had been washed away the night before, my dad

An aerial view of Sea Isle City, New Jersey. The bridge to Avalon can be seen on the left. The Polaris Motel was undermined, and its pool was washed away. Oceanfront homes to Ninetieth Street were damaged. *Courtesy of Sea Isle City Historical Society.*

borrowed a ladder from his good friend Lonnie Peterson so they could access the first floor of the house. Upon arriving at our home, he soon realized that the ladder was unnecessary.

All that was left where our home stood was a mangled pile of rubble. Our family had escaped the storm with our lives—but literally with only the clothes on our backs. My mother, my two younger brothers and I were eventually evacuated by helicopter from Sea Isle City and went to live with my oldest sister and her husband in Bordentown. My father, older brother, grandfather, uncle and Runzel, the wonder dog, all stayed behind to begin rebuilding our ravaged town.

By the grace of God, we all had escaped with our lives and had an unforgettable storm story where our family and neighbors had to "run for their lives." One of the worst winter storms of the twentieth century.

Margaret Thomas Buchholz

Margaret Thomas Buchholz is the coauthor of *Great Storms of the Jersey Shore* and a recognized, accomplished author, having written several acclaimed publications, all centered on her long lifetime on the New Jersey Shore. A graduate of Cedar Crest College, she and her husband purchased *The Beachcomber* in 1955, a weekly summer publication highlighting the charm and enchantment of everyday life on Long Beach Island. It remained in publication until 1987.

> *When the March storm swept over Harvey Cedars, I was in Philadelphia. My father, Reynold Thomas, was mayor. He had been evacuated by helicopter in order to meet state and federal officials on the mainland. I drove down to Manahawkin, where I met him. A state policeman drove us back onto the island and dropped us off at the south end of town, where the visible road ended and we had to walk on the soggy sand through town.*
>
> *The ocean had washed over the island to the bay, destroying almost all homes within twelve blocks. As the enormity of the damage became visible, my father began to cry. I had seen him cry only once, when my mother died three years earlier. Four houses were still standing on the ocean side, perched on pilings with the tide washing under them. The land was totally flat from ocean to bay, no dunes, no greenery, no main road and only one house still intact on the bayfront.*
>
> *Farther along what had been the main road, an Atlantic City Electric Co. truck was overturned in a puddle of water, near the barely recognizable piece of a house and a flattened electric pole with seaweed wrapped around the wires. As we came into town, working our way around chunks of buildings and tottering telephone poles, we passed the borough hall, split like a coconut, and approached Seventy-Ninth Street and stopped. The power of the waves created a new inlet, and the water flowed from the ocean to bay. Within twenty-four hours, my brother had moved the family dredge into position in the bay and began filling the break. (My father owned Barnegat Bay Dredging Co.)*
>
> *After four days and nights, the federal dredge came and finished the job. A week later, a newspaper account reported: "Nobody lives in Harvey Cedars anymore. There are no roads. There is no drinking water. The permanent residents have been evacuated and where there is no wreckage there is just sand, in some places four feet deep. Even the houses don't offer evidence of where the roads were. The houses are everywhere, in no order, sometimes*

> *piled two or three together. Around them crushed and mangled cars and trucks lie half buried."*
>
> *My father, a former marine, promised, "We'll be back to normal soon." Pointing to an American flag flying from an almost totally destroyed house, he added, "I guess that will show you that we intend to win this battle."*

Joe LaRosa

Joe LaRosa, who was just a boy when the storm hit, still sees every terrifying moment of the Great Atlantic Storm of 1962. A native of Sea Isle City, he has boyhood memories of the charming beach community, playing basketball, spending days on the beach, playing miniature golf and gathering with family in the summer. He lived at Forty-Third Street and Central Avenue on the barrier island. He was almost nine years old as the second weekend in March 1962 held a meteorological secret that no one suspected.

LaRosa worked in public relations for Cape May County and in hotel hospitality in Atlantic City. He is an educator, a teacher and principal of Dennis Township Elementary School. He is a people person, and he has witnessed the Ash Wednesday storm through the lens of storm survivors who endured the fury of that savage storm sixty-two years ago.

He has authored two books about the Ash Wednesday storm, *Our Perfect Storm*, released in 2010, and *Storm Stories*, published in 2019. Now, he still has a story to tell.

> *"I thought Hurricane Donna was bad, but this looked like an atomic explosion compared to it," said Captain Paul McGill, an Eastern Airlines pilot, after flying over the New Jersey Shore.*
>
> *One overriding thing you have to understand, flood insurance was so cost-prohibitive that no one had flood insurance. If you lost your house, you lost your house. If you lost your equipment to work, you lost your work. If you lost your tools, you lost your tools. There was no such thing as FEMA. The government did not come in and assist all of us. What we relied on was each other.*
>
> *Self-reliance as well, that is what built the town back up. We had no help, and the government—President Carter started FEMA seventeen years after the '62 storm. People ask me about the storm to this day. My God, it was like a hurricane. We had a nor'easter recently; it was pretty bad. The '62 storm was ten times worse. A low-pressure system was coming*

across the United States. It was coming from the Midwest, dropped down south and went up the Shenandoah Valley. There was record snowfall in the Shenandoah Valley in western Virginia. It snowed so hard that a snowplow train got stuck, and they couldn't move the snow at all; that was a very rural area.

Down in Florida and the Carolinas and Georgia, there was another low-pressure system, and it merged with the Midwest storm over Maryland and Delaware, merged into one powerful storm. There was a giant high-pressure system stretching from the Arctic into eastern Canada and all the way over the ocean to Iceland. The coastal storm had nowhere to go, it was stuck over the Mid-Atlantic. It was the perfect storm, there was a perigean moon, which occurs when the moon is at its closest point to the Earth, and this compounded the effect of the tides on the coastline.

There were six high tide cycles, although the tides never really went out, for three days, nonstop. Remember, there were no satellites for weather. We didn't have any warning about the storm. The day before was a beautiful day; it got cloudy by evening. Before dinnertime, my brother Pat and his friends Angelo and Dewey were playing basketball. It started sleeting and raining. We had been through so many of these nor'easters before, there was no big deal.

Now the next day was Wednesday, March 7, and a couple of kids I knew were altar boys, and they couldn't get to the church to get the ashes. Then the first high tide came up; it was a very heavy tide, and it badly damaged the beachfront and the first house was knocked down. On Monday evening, while the three aforementioned boys were playing basketball, one block down on Forty-First Street Parkway at the Veterans of Foreign Wars building, Bartholomew "Bart" Milano, a Sea Isle City resident, was playing cards with other veterans.

After finishing his game, he left his friends and went home to his house on Fifty-Seventh Street. His house was the last one on the block facing the ocean and was over two stories high with a large covered porch facing the beach. The two-story building started above ground level and was built upon pilings that raised the house high enough to provide protection from the infrequent above-average high tides while also conveniently serving as garage space for his automobile.

As he parked his car, he noticed that the wind was picking up in a strong northeasterly direction. Having ridden out many of these late winter storms, he parked his car under the house and went inside, where his wife, Mary, and three boys, aged eleven, three and two, were sleeping and retired for the evening. Before dawn, he was awakened by a pounding noise from outside of his bedroom window. When he arose and looked out, facing the ocean, he was startled to see that the roof section covering his porch was no longer attached to the house.

The ocean was up to his home and pounding the foundation wall. He immediately woke his wife and told her to get the children ready to leave the house. As the wind and ocean continued to attack the building, he quickly dressed and went outside to prepare the automobile to leave. He found his car in water over the door jamb. As the car started, he eased it out from under the house by feathering the clutch so water would not enter through the exhaust system.

As the car backed out from the concrete pad and entered the driveway, it promptly became stuck in the high water and mud. Realizing that his family would need to escape on foot, he reentered the house to gather his wife and children. Two weeks prior, he had just completed a renovation to the back door of his home, replacing a large door with a slightly smaller one

Houses were picked up and deposited during the Great Atlantic Storm of March 5–9, 1962. Twenty-Eighth and Landis Streets, Sea Isle City, New Jersey. *Courtesy of Sea Isle City Historical Society.*

with two glass panels on each side. As his family tried to leave the house, the new door would not budge. The pounding of the surf had dislodged the house in such a way as to shut the door tightly. In order to escape the building, Bart kicked out the door.

He cursed to himself while he kicked the door open, because in opening it, he broke one of the side panels [he had] *recently installed. He thought to himself that he now would need to repair the door when he returned to the house when the tide receded. The family left the building and made their way to the home of Bill McClory, located two houses away on Fifty-Seventh Street and Pleasure Avenue. Safe and warm, the family realized that they had left their pet dog back in the house. Bart returned by wading through the cold, high water to his house in order to rescue the animal and bring it to safety. He went up the back steps of his house and called for the dog. After coaxing the dog to him, he grabbed the animal and fled down the back stairs.*

As he reached the second step from the bottom, he felt the structure pull away from him. Turning back, he saw his house with everything he and his family owned falling—almost in slow motion—into the sea. The entire house fell into the sea, crushing his car beneath it. As the debris fell from the house, some became lodged in the horn of the car, causing it to blow non-stop until it was also overpowered by the ocean.

The Milano family was the first victim of Sea Isle City's perfect storm.

Alice LaRosa didn't bother to wake her son Pat, one of the boys who played basketball the night before. There would be no Wildwood Catholic High School entrance examination for him this morning. The early morning tide had been rising and had shown no indication of receding, and anyway, she was unable to get the family automobile off of the island. By 7:00 a.m. the LaRosa house at Forty-Third Street and Central Avenue was completely isolated, as the tide from the bay met the tide washing down Forty-Third Street from the beach. As far as could be seen from the house, both Forty-Third Street and Central Avenue were completely flooded.

In addition to Alice and her son Pat, also in the house was the family's youngest boy Joseph, aged nine. Joseph shared his name with his father, who was currently at work serving as the assistant superintendent of public works in Sea Isle City. The strong northeast wind had started to send debris from the oceanfront down Central Avenue. When the tide grew higher and the wind kept blowing, the family house first lost telephone service and then electricity. As their mother tried to maintain her composure, the two boys laughed when they looked out the side window of their home and saw a small bungalow house floating down Central

Destruction along the beachfront, with homes destroyed, following the second and third catastrophic high tides during the Great Atlantic Storm of March 1962. *Courtesy of Sea Isle City Historical Society.*

Avenue from Forty-Second Street. The house still had its front porch attached, with rocking chairs still rocking and curtains and lamps clearly visible in the cottage's windows.

By now, Chief Rochester was well aware that this was a very different storm. At 4:00 a.m., the barometric pressure at the station had bottomed out at 29.34 inches of mercury, and winds were blowing steadily from the north at a whole gale, between 55 and 63 miles an hour. At 8:00 [a.m.], *the barometer rose slightly to 29.50 inches, but the winds were still at a whole gale. Compounding the situation was the morning's high tide that arrived slightly after 7:00 a.m. at the beachfront.*

It would not arrive in the back bays until later in the morning. By 9:40 [a.m.], *the officer of the day recorded that the station was completely flooded under four feet of water and the bathhouse was submerged under more than five feet. As the wind kept blowing, the seawater kept rising. In 1962, Sea Isle City was a very small, sparsely developed community. Indeed, there was very little building from the area near the new shopping center on Sixty-Third Street until the beginning of Townsend's Inlet at approximately Eightieth Street.*

A destroyed carousel, Boardwalk and Forty-Third Streets, Sea Isle City, New Jersey, after the great Atlantic storm of March 1962. *Courtesy of Sea Isle City Historical Society.*

Many of the numbered streets were not yet cut between Landis and Pleasure Avenues. Central Avenue was also very lightly developed. The United States Coast Guard (and prior to that, the United States Lifesaving Service) has had a presence in Townsend's Inlet since 1849. [Of the] *men assigned to Coast Guard Station #130 (and prior to that United States Life Saving Station #34) six have always assisted Sea Isle City during emergencies. This storm would be no different.*

At 4:15 p.m., Chief Rochester received a notification that a fire had broken at Mercy Hospital. The Meadows got flooded, completely filled up with water. So here comes the second tide that night. Our house was shaking. Outside, the water pipes broke, and the water rushed out into the ocean; sewage was just gushing out of the sewage plant. The phone system failed. Remember, this was 1962—no cellphones. Our electric got knocked out; thankfully, some places in town had electricity. My dad was assistant superintendent of public works. Sea Isle City was a very *small town, and it was a different era.*

When I was off from school (he also worked half days on Saturday), I'd drive around town in the public works truck with him while he did his chores. That is how I became very familiar with the town and boardwalk and residents at a very young age. On the first night of the storm, my brother and I went to our bedroom to try to sleep. Only my mom was home, and it was pitch black out. We had no electricity; it was cold, and the wind was howling. Our house was at the northeast corner of Forty-Third Street and Central Avenue, and our bedroom was on the third floor. The house would actually shake when the big gusts hit. Sometime later, we were called down to the living quarters. My dad was out helping the community, and he had brought a city truck home.

Our entire house was surrounded by at least three feet of water, so he pulled the dump truck across our backyard and parked it near the back stairs. It was facing forward. We went down the back steps and pretty much jumped from the steps into the truck and then over to the passenger side. It was cold, rainy, windy and very dark. Once we were in the truck, my dad drove us the four blocks to my aunt and uncle's home. There was my dad and mom, me and my older brother. We were the only people there except for my aunt and uncle. All three of my cousins were away at college, so we all had our own beds. Being at the house was no big deal, as we would go there very often. When I awoke it was after daybreak, I went downstairs, and people had been coming to the house and it was quite full.

There were probably thirty people there. My uncle Dan and aunt Grace started planning our escape. Houses were cold, dark, no heat, no electric, no phone, no nothing. He opened his doors up to the community. And as I said, we were starting to think that this was different from other storms. Also, during the first day, just focus on things. Then there was a fire at the hospital. There were lot of fires on the barrier islands during the storm. Usually, the chief would dispatch a crew to go help fight fires or go direct traffic.

How did the fire get reported? By someone running through the water to get the word out to Chief Rochester that there was a fire at the hospital. We had a fifty-four-bed hospital, and he dispatched a crew. Well, their vehicle gets swamped around Seventy-Seventh Street. They walked from Seventy-Seventh Street to Fifty-Ninth Street in thirty-five-degree water. They get there, and what happened is one of the maintenance men took an axe and he cut out everything that was on fire until the fire services arrived.

Chief Rochester immediately dispatched a squad of five men to assist the Sea Isle City Volunteer Fire Company in fighting the fire. The group left the station at Eighty-Second Street in an effort to reach the hospital located at Fifty-Ninth Street and Landis Avenue. By 5:00 p.m., as they neared Sixty-Sixth Street and Landis Avenue, their vehicle was inundated by the strong surging tides and downed power lines. They could take their vehicle no farther, so the men continued with their mission on foot, wading through the cold, swift waters.

The destroyed boardwalk at Sea Isle City, Great Atlantic Storm of March 5–9, 1962. *Courtesy of Sea Isle City Historical Society.*

By 7:30 p.m., the squad reached the hospital. Although the fire had been extinguished by a maintenance worker, there were still many ill people at the facility that needed to be saved from the rising waters. The coast guardsmen immediately started to assist hospital employees. At 10:00 p.m., the men received orders that they were to remain at the hospital and help as needed. Chief Rochester's day was not finished, as earlier in the evening he decided to move the station's cutter from the open intercoastal waterway near the inlet to safe harbor in the center of the island. Coast Guard Cutter 30432 was moved to a safe haven on Venetian Road, at the site of the old Dever's Boat Yard.

Indeed, the fire at the hospital was a great concern to the city's fire department. A structure fire, being whipped by the gale-force winds associated with this storm could conceivably engulf the entire island. After surveying the initial damage after the first high tide, Fire Chief Willard Wright ordered the town's firefighting equipment dispersed throughout the island.

In March 1962, weather forecasting was still an inexact science. Advances in forecasting methods spurred along by World Wars I and II had brought meteorology into a new era, where forecasts were based on scientific methods and were made with a sense of reliability. The world's first weather satellite, TIROS (television infrared observation satellite) had been launched not even two years prior, but information from the satellite was not available to forecasters. Indeed, the satellite was on a polar orbit and served to only map cloud cover of that area. There was no weather satellite to observe what was happening over the eastern United States.

Although data interpretation had grown with scientific advances, the actual collection of weather reports had changed little since the beginning of the United States Weather Bureau in the 1880s. At that time, the Weather Bureau started as part of the United States Army Signal Corps because that was the only organization that could communicate effectively over large areas. The heart of the weather service and the thing that had to exist before there could be such service was the telegraph. It allowed, for the first time in history, the rapid, simultaneous transmission of information from stations thousands of miles away.

The first systematic collection of weather data producing an actual weather map was placed on public display in Washington, D.C., in 1856. At that time, the Smithsonian Institute collected weather reports from over three hundred volunteers from around the county and produced the map, giving visitors a snapshot of weather conditions from across the county. In reality, before the era of weather satellites, weather forecasting was, in a very crude sense, simply contacting someone to the west of you and asking them to look outside their window.

Weather in New Jersey moves from west to east, and the prevailing winds over the state are called the "westerly airstream." New Jersey is located in the core of the westerly airstream, and because of that, it experiences many different types of weather. On any given day, weather sources may include the arctic tundra of Canada and Alaska or the cool waters of the North Pacific. The weather may originate in the hot desert sands of the Sahara.

It may have its origins in the Southwest and Mexico, the warm waters of the Gulf of Mexico, the Caribbean Sea and the adjacent tropical Atlantic, or even in the chill of the maritime environment of the North Atlantic, east of Canada.

In March 1962, two major weather systems coming from two different western areas were converging on Sea Isle City, New Jersey. As my brother Pat and his friends Dewey and Angelo were playing basketball in the light mixture of sleet and rain of March 5, 1962, without their knowledge, weather forces were lining up to create Sea Isle City's Perfect Storm.

Sea Isle City and the entire South Jersey area was served by television stations from Philadelphia; in March 1962, only three television stations were available for viewing at the shore. They were WRCV, Channel 3, an NBC affiliate; WFIL, Channel 6, an ABC affiliate; and WCAU, Channel 10, an affiliate of CBS. Most people tuned into the evening news from one of these stations to get the daily weather forecast. On Channel 3, the weatherman was Wally Kinnan. Kinnan was known throughout the broadcast area as "Wally Kinnan the Weatherman" and had been the first weatherman in the Philadelphia area to give a "five-day forecast."

On Channel 6, the weatherman was Dr. Francis Davis. Dr. Davis was a trained mathematician with a degree from Drexel University in Philadelphia, and he had learned meteorology while serving in the army during World War II. During that time, he had served on the meteorological team that gave the famous weather forecast that sent the Allied forces into Normandy on June 6, 1944. He later earned his PhD from New York University and was a certified meteorologist. He had been with WFIL since the late 1940s and made the transition from radio into television when Channel 6 went on the air.

The weatherman on WCAU, Channel 10, was Herb Clarke. A native of North Carolina, he had a degree from Bowling Green University and was a navy veteran of both World War II and the Korean Conflict. Herb had been giving the weather at WCAU since 1958. Already very popular with the viewing public, he would go on to have an illustrious career at Channel 10, reporting the weather until 1997.

All of the weathermen were local celebrities and provided the most up-to-date weather forecasts available. These forecasts were generated at the United States Weather Bureau office at the Philadelphia International Airport. In print, the local daily weather forecasts were also given by the area's daily newspaper, The Atlantic City Press. *The Weather Bureau offices in Atlantic City generated this forecast. During the past few days, the weathermen at the Philadelphia television stations were tracking a severe storm across the United States and offered a warning for the possibility of high tides in the South Jersey Shore viewing area.*

The Philadelphia National Weather Bureau office provided these forecasts. The Weather Bureau had limited space at the airport and was by no means a large governmental organization. The office had a staff of seven people, including their secretary, and served the Philadelphia metropolitan area—the third-largest population center in the United States.

Additionally, with the exception of a thermometer and perhaps a barometer, the television stations had no forecasting tools of their own.

They did not even have a media teletype from the National Weather Service to warn of imminent emergencies. The forecasts were received over the telephone by the television stations immediately prior to the 6:00 p.m. and 11:00 p.m. news broadcasts. They were then aired throughout the Delaware Valley. Although these weathermen were using the latest technology in making their predictions, their manpower and forecasting tools were limited and primitive by today's standards.

On Monday, March 5, The Atlantic City Press *ran a story on its front page describing a major winter storm in the Midwest. Huge amounts of snowfall along with high winds stranded entire areas of the Dakotas and dumped twenty-six inches of snow in Minneapolis. The heavy snowfall completely paralyzed the interior of the country. In a local interest story,* The Atlantic City Press *also reported that a live seal washed ashore in neighboring Avalon and was taken to the Philadelphia Zoo by the local police department in an effort to save its life.*

There was no indication about the storm to come.

As the boys' basketball game continued, the ominous storm moved closer to the Jersey Shore and to Sea Isle City. By the Tuesday, March 6 edition of The Atlantic City Press, *storm reports were being received by areas closer to the South Jersey Shore. The newspaper warned local residents to be prepared for snow inland and rain and snow at the shore. It predicted strong winds—forty to fifty miles an hour—and tides as high as one and half feet above normal. Winds were predicted to start off in a northeasterly direction and then change to a northwesterly flow.*

A newswire story from the Associated Press was also printed in the newspaper and described cold air streaming from the Midwest bringing snow to the Deep South. The snowfall was so heavy that two Chicago and Northwestern railway snowplow engines were stuck in wind-blown drifts in Iowa. In the South, widespread snow fell in North Carolina, Tennessee and northern Georgia. Snow had fallen as far south as Columbus and Greenville in Mississippi.

Frost and freeze warnings were issued well into the interior sections of Texas and Alabama, and a heavy snow warning was issued for western and central Virginia and northern Maryland. For the New Jersey Shore, the National Weather Service was predicting a winter "northeaster" storm, and the television stations and newspapers had picked up the story. The shore was no stranger to these winter storms. It had withstood many severe gales.

What was unknown to most people was that weather and astronomical forces were lining up to make this storm significantly different from all of the other "northeasters" that had hit Sea Isle City since it was settled in the 1880s. The shore was experiencing a "perigean spring tide." This tide is the highest average tide in the solar-lunar cycle, and it occurs when the moon is closest to the earth while in its orbit. The effect of the moon on the Earth's tides is significant. During this time, tides run at their highest volume.

In addition to the high tides, a very large and very strong high-pressure system stretched over the northern tier of North America, all the way across the Atlantic Ocean and into England. The east-to-west stretch of this Canadian high was incredibly long and would soon act as a giant roadblock, stopping any other weather systems in its path. While this high-pressure band was in place, an intense winter storm was bearing down on Sea Isle City.

National Weather Service forecast for Atlantic City and vicinity for March 6, 1962
From: The Atlantic City Press

> Quite windy today. Rain mixed with some snow along shore areas, but mostly snow in the interior regions. Chance of rain becoming mostly snow later today and continuing into tonight. Highest temperatures in the 30s today and tomorrow. Outlook for Wednesday, cloudy and cold. Tides will run 1½ to 2 feet above normal with minor flooding at times of high tide today.

Due to the high, surging tides, movement between areas of the resort was impossible, so firefighting equipment was strategically placed throughout the island. This move would enable at least one piece of equipment to respond to most areas in case of an alarm. One engine was placed in Townsend's Inlet, two pieces of equipment were placed at Davies' Sunoco station between Forty-Ninth and Fiftieth Streets and Landis Avenue, and another engine was placed at Morano's Atlantic gas station at Forty-First Street and Landis Avenue.

Chief Wright's prediction came true, for when the alarm was called at the hospital, the fire department could not get their equipment past Fifty-Third Street and Pleasure Avenue, as all other avenues were completely submerged under very deep water or completely covered by debris that made

A beachfront home collapsed after five catastrophic high tides during the Great Atlantic Storm of March 5–9, 1962. *Courtesy of Sea Isle City Historical Society.*

the streets impassable. Responding to this alarm, the firemen also finally reached the hospital on foot.

City Superintendent Howard Wright was a lifelong resident of Sea Isle City. He was in charge of all water, sewer and roads in the community. This high tide was much different than those he had experienced before. By mid-day, the city's potable water and sewer system were overwhelmed. In an effort to get some of the system working he went to the West Jersey Avenue home of Jim Coulter in an effort to have the electrician get some of the city's pumps working.

The seawater was rising up the stairs inside of the Coulters' apartment, but as a member of this close-knit community, Jim felt that he had to leave his family and offer any assistance that he could during this emergency. Jim made his way to the garage, where he kept the equipment not in his truck, gathered what tools he could and went to the city's sewage treatment plant at Forty-Seventh Street and Central Avenue. Somehow, he got the pump working. The attempt was futile, as the storm was not ending—it was only beginning. Eventually, the city's water and sewer systems were overwhelmed. They remained inoperable for weeks after the storm finally ended.

Meanwhile, seven blocks up-beach north of Milano's house, George "Pat" Haffert was happy. His house, which his family named "Surfside," as it once served as a hotel, was located by the ocean at 5006 Marine Place—directly on the beach, accessible from the street by stairs from their backyard. The sound of waves pounding against his house could only mean one thing, and that was no school today! Although stormwater hitting the house was not totally uncommon, the waves today seemed a little more threatening.

Indeed, his father, Horace, had mentioned that he expected a very high tide and even the potential of a big "cataclysmic" event. The Haffert family were the owners of Sea Isle City's largest employer, the Garden State Publishing Company. With grandparents, uncles, aunts and cousins, the Hafferts were a very large, extended family in Sea Isle City. As the day wore on, Pat's father, Horace, invited his parents and his brother Bill and his wife and their children to their oceanfront home to "ride out the storm."

While they dined, the tides continued to pound the house, with water now knocking at the door facing the ocean. Concerned, Bill got a two-by-four board and nailed it against the door to provide support. This storm was starting to feel like none other. Just as Pat's mother, Helen, placed dinner on the table, the family felt the entire building lurch forward. Everyone was thrown to the floor, cabinets opened and glasses, plates and food went flying. That terrible lurch meant only one thing—the pilings supporting the house were being undermined by the strong tidal surges and the house, with them inside. [The house] *was in imminent danger of falling into the sea.*

With the lurch, Horace yelled to his son Joseph, "Take the children and run!" Herding the children together, Joe led his brothers Pat, Mark and Greg; his cousins Mike and Jan; and the family dogs out of the house. Much like the Milanos, Pat's cousin Mike needed to break the glass in the door to escape the house. Cutting through backyards, they eventually met up with the adults of the family and made their way to Lewis's Funeral Home, located on Fifty-Third Street between Landis and Central Avenues, where they were to spend the night. Adding to the macabre scene that they had just lived through was the presence of a corpse in the parlor's viewing area. Services for a deceased resident were scheduled for that evening. The services were postponed.

Seven-year-old Lien Dalrymple's shoes were wrecked, but that was the least of his parents' concerns. Immediately prior to the morning's high tide, Lien's father, Charles, was attempting to salvage as much stock as possible from the family store on the boardwalk. The "Penny-Lien" was an apparel shop located on the boardwalk at Forty-First Street near the doughnut bar and bowling alley.

An aerial view of beachfront and boardwalk destruction following five catastrophic high tide cycles, Great Atlantic Storm of March 5–9, 1962. *Courtesy of Sea Isle City Historical Society.*

The high surf was pounding it, and the family was in danger of losing all of the stock now in place for the upcoming summer season. Charles, who was also a teacher at the Sea Isle City Public School, was taking stock from the shop on the boardwalk and moving it to his home, located one-half block from the beach.

The top story of his house gave a very clear view of the boardwalk, the movie house on the pier and the raging surf. Meanwhile, outside his front door, the ocean was roaring down Pleasure Avenue in a rough current, bringing debris from the beachfront by their house. They were marooned in the house until the tide subsided.

As bad as the situation was at the "Penny-Lien," things were much worse at the Madeline Movie Theater that was on the ocean side of the boardwalk. The Madeline Theater had lost its stage area and screen during the previous day's tides. It was now being repeatedly battered, as each wave to hit the structure seemed higher than the next. Along with other family members, Charles had entered the structure in an effort to salvage what could be carried out. While in the movie house, he could feel the pilings that supported the structure repeatedly being jarred and moved as the waves crashed.

It was decided that it was no longer safe to remain in the building, and he went home to tell his wife and children about the storm. He also told the children to go to the top story of their house and look at the pier. The children were not at the window watching for more than a few minutes before they witnessed the entire pier and movie house crash into the raging surf. It was totally swallowed by the sea. Later that afternoon, while the tide was dropping, Lien and his family were able to leave their house safely and go to a friend's house on Forty-Fifth Street that still had electrical service.

They stayed at the house until the next morning. There was twenty-four hours of driving rain, gale-force winds and surging, uncontrollable tides that were taking their toll on Sea Isle City. Successive tides were pushing seawater onto the beachfront and into the wetlands that separated Sea Isle City from the mainland. The high tides and wind kept the water in the wetlands, and each successive tide added more water until almost everything on the island was covered in water. The seawater in the bay areas could not escape back into the ocean, so it backed into the city. Sea Isle City was caught and being squeezed in a wet and terrible vise. The oceanfront continued to receive the pounding surf, while seawater with no escape route to the ocean inundated the island from the bayside.

Sister Marie Fidalis was principal of St. Joseph's Elementary School in Sea Isle City. A member of the Merion Sisters of Mercy, she first came to Sea Isle City in August 1962, immediately after the March storm. She transferred from Sea Isle City in June 1968. In addition to damage to St. Joseph's School during the March 1962 storm, the order lost three convents. One needed to be evacuated during the first high tide and was lost to the sea soon afterward.

Many of the sisters who assisted at Mercy Hospital were housed there and lost all of their belongings. The destruction was total, as the buildings "collapsed like match boxes." One of the sisters who was evacuated was our Mother Superior Sister Pierre. The school sustained a great deal of damage. As you entered, the high-water mark was obvious throughout the entire building. The flood undermined the foundation, and that needed to be repaired to keep the school open. All of the furniture and desks needed to be cleaned. As a result of the storm, the convent and church needed significant repairs.

In March 1962, Mike Stafford was a young teacher at the Upper Township Elementary School. He has a very long association with Sea Isle City, including writing books about the town that he loves. They include Sea Isle City *(Images of America),* Brides of Sea Isle City and Beyond *and* I'd Do It All Again...But Slower. *He is the president and guiding hand behind The Sea Isle City Historical Society and Museum. At this house at Sixtieth Street and Landis Avenue, the tide was up to his car on the morning of March 6, 1962, and he knew he needed to get to work. As he tried to get offshore, his automobile stalled in the water, and he was towed out by Mr. Mike Davies, son of Joseph Davies, the city fire chief and proprietor of Davies' Garage on Landis Avenue. Getting back to his home, he knew that this was a very different storm. He had heard that the Milano family had just lost their home on Fifty-Seventh Street to the wrath of the ocean.*

As the storm raged on and more homes became damaged, Mike took people who were without electric or heat into his home. Miraculously, throughout the entire ordeal, his home did not lose electric or telephone service. Between tides, Mike explored the island going to the beachfront in time to witness the Sisters of Mercy Convent being lost to the sea at Sixtieth Street and the beach.

When we returned from the evacuation, he came home to a city that was "united" and "did not lose spirit." Preparations for the 1962 summer season spawned many great ideas, including the eventual "Skimmer Weekend" that is still very popular every June.

Harry Bellangy

Harry Bellangy is a retired computer system architect and has served as president and historian of the Greater Cape May Historical Society for twenty-five years. The society was established in August 1974 and is headquartered at the Colonial House Museum (built circa 1730). This house was originally a tavern and the family home of Revolutionary War Patriot Memucan Hughes. Bellangy is a lifelong native of the town and is deeply committed to preserving and sharing the rich history of Cape May.

Cape May sustained substantial damage during the Great Atlantic Storm of 1962 yet escaped the catastrophic damage that occurred farther north at Sea Isle City on Ludlam Island and Long Beach Island.

We did not have a seawall in 1962 and thus had no protection from the ocean. All that was there were pilings, and they provided little protection. The storm beat us up for four days, the five high tides, and the upward wave action devastated the beachfront. The fifteen-block boardwalk was ripped up, splintered. Convention Hall, a Cape May landmark, was destroyed. We were fortunate there was no loss of life. Two years later, in 1964, the seawall was built. It runs along Beach Avenue, not far from the water tower, which holds 736,000 gallons, which is on Columbia Avenue.

A house submerged at Sixtieth Street and Seventh Avenue, Sea Isle City, after Great Atlantic Storm of March 5–9, 1962. *Courtesy of Sea Isle City Historical Society.*

I was away at Glassboro College during the storm. I could not get into Cape May due to the storm; I had to go to Ocean City and stay there until conditions improved. The high tides drove the water far into town; the old Beach Theatre had three feet of muddy sand inside. The ocean water reached all the way to Franklin and Hughes Street, where I lived. That was a half mile; there was nothing to stop it.

Cape May has a lot of history critical to New Jersey. There was a large industrial plant, the Harbison Walker Cape May Works, which processed water from Cape May Bay to extract magnesite from the seawater to create refractory bricks, which were a critical component in furnaces and boilers during World War II. The plant was undamaged during the storm. There were also the "ghost tracks," railroad tracks over a century old, which often appear after big storms, only to disappear under the shifting sands.

The town was able to recover, at least partially, in time for the 1962 summer season. A lot of work was done in the ten weeks from the storm until Memorial Day. At least we had a summer. I worked for the amusement arcade, which lost eighteen Skee-Ball games during the storm. Most businesses survived, although Tenenbaums Arcade and Morrow's Nut House reopened in a new location.

4

THE LINDSAY SNOWSTORM

FEBRUARY 9, 1969

Of all the winter storms that have battered New Jersey and ground the Garden State to a standstill, there are only a few that can be characterized as transformational. Their impacts were so profound and enduring that they reshaped storm mobilization, transportation, infrastructure, preparation efforts and even the politics of New Jersey's municipal and state government. One was the Great White Hurricane, the Blizzard of 1888, and over a century later, there was the Blizzard of 1996.

Then there was the Lindsay snowstorm!

There are still countless New Jerseyans and New York City and Long Island residents who remember waking up that Sunday morning, February 9, 1969, astonished at the relentless snowstorm that was pounding the region. Children were delighted, but adults were fighting a losing battle to clear their walks and driveways to liberate their automobiles. The driving snow, falling at the rate of an inch per hour from 3:00 a.m. to 5:00 p.m. then accumulating at lighter intensities, fell until the final flakes drifted down at 1:00 a.m. Monday morning.

Although it was not a blizzard by definition, the storm had gale-force winds and heavy snow that resulted in borderline blizzard conditions. This was a winter storm with many stories, the biggest one being the near total absence of accurate forecast information that could have afforded New Jersey cities and townships, like Paterson, Newark, Elizabeth, Edison and Trenton, as well as many other smaller municipalities and towns, precious time to anticipate the intensifying nor'easter that was moving northward just off of the Eastern Seaboard.

The storm was forecast by most television and radio media. A significant storm was anticipated. However, a snowstorm was not predicted! Almost all forecasts called for snow developing after midnight Sunday, quickly changing to heavy rain that was to continue until Sunday night and possibly changing back to snow briefly before ending. Street and highway flooding was expected. There was no mention of snow accumulations for North and Central New Jersey, except for up to four inches in the far-northern and northwestern portions of the state.

A few forecasters hinted at a colder scenario and significant snow. CBS meteorologist Gordon Barnes told radio listeners Saturday that the nor'easter might move just far east enough for the region to see rain changing to accumulating snow. ABC weatherman Tex Antoine told viewers that he might have to issue apologies on the Monday evening telecast.

By late Saturday night, the low-pressure area redeveloped and intensified dramatically, farther east than forecast, and heavy snow enveloped north-central and Northern New Jersey and New York City, beginning briefly as rain. Fifteen inches fell at Newark, twenty inches at Paterson and over two feet in northern Bergen County. Thirty-mile-per-hour wind gusts resulted in three-foot drifts, and on the New Jersey Turnpike, barely a car could be seen.

Damages were estimated to cost millions, with the storm claiming ninety-four lives with over three hundred injuries, mostly in New Jersey and New York City, where forty-two people perished. A joint session of the New Jersey state legislature was postponed, schools were closed and mass transit was shut down. The unexpected snowstorm immobilized cities from Trenton northward, closing schools for days and leaving millions of Jerseyans, who did not expect fifteen to twenty inches of snow, housebound, struggling to dig out. In New York City, it was a nightmare; 40 percent of the city's snowplows were disabled and being repaired, in no condition to move tons of wet snow. Twenty-one people died in Queens alone, and snowplows took a full week to reach the outer borough. Cabs were nonexistent, and garbage piled up by the day.

New Yorkers turned their vengeance on Mayor Lindsay, who was accused of giving Manhattan snow removal priority. Known to this day as the "Lindsay snowstorm," it resulted in Mayor Lindsay's upset defeat in the New York City's June Republican mayoral primary. The infamous snowstorm forever altered how New Jersey local governments prepared for approaching storms, regardless of the season. There was collaboration with the National Weather Service and local forecast agencies, which led to permanent improvements in storm and public safety preparations.

George Friedman

George Friedman is a publisher and editor-in-chief for the *Securities Arbitration Alert*, a weekly online publication that covers the latest developments in financial services arbitration and mediation. A native New Yorker, he and his wife, Ellen, have lived in Teaneck for forty-eight years. He is an accomplished writer, a retired adjunct professor of law at Fordham Law School and a lecturer on the subject of alternative dispute resolution.

I was reminded that we just celebrated the forty-fifth anniversary of the great snowstorm of February 9, 1969, also known as—for reasons that will be explained later—the Lindsay storm. As bad as it has been around here this winter, that storm was worse. Much worse….

When I was a kid (I was fifteen in 1969), I loved the weather, especially snowstorms. I haven't changed as I close in on my seventieth birthday. Problem was, I lived in Douglaston, in northeast Queens, New York. The proximity to the relatively milder Atlantic Ocean meant that many an expected snowstorm became a rainstorm. I hated that and often wished that—just once—the dreaded rain/snow line would halt its inexorable northward march just south of Douglaston. I later learned you have to be careful what you ask for.

The forecast was for snow rapidly mixing with and changing to rain, with highway and urban flooding. My friend Gene—another weather guy—and I were disappointed and annoyed as usual. I should point out that Gene is my only childhood friend who actually grew up to become what he said he would become: a weather forecaster. He retired a few years ago after a long career at the National Weather Service and was and is the best forecaster ever. I planned on becoming an astronaut but—spoiler alert—that didn't happen. He's also the only childhood friend with whom I've stayed in close contact. Yup, those weather bonds are durable. Even though he's on the West Coast, we still get on the phone to talk about approaching storms. Gene saw something in the weather maps most professionals didn't. The day before the storm was to hit, he said, "I'm not so convinced it's going to turn to rain. If that's the case, watch out!" WABC weatherman "Tex" Antoine had similar reservations and doubts. Said he on Saturday night (the storm started Sunday morning), "It may be dicey north of 125th Street. I'll be back with my apologies on Monday."

Rain? What rain? Sunday dawned cloudy and gray. It started sleeting and raining at daybreak, which annoyed me greatly. But then, miracle of miracles, the rain/snow line shifted south, and it turned to snow. Lots of snow. Hours and hours and hours of wind-driven snow. The city (that's what New Yorkers call Manhattan) officially recorded 15 inches of snow. Not Douglaston. It is the highest natural point in New York City—about 260 feet above sea level and is also the easternmost point in New York City. That combination gave us—ready—30 inches of snow! That's right, 2.5 feet. The drifts were incredible, 6 feet in some places.

The aftermath: Lindsay's folly.

Somebody evidently forgot to tell then-Mayor John V. Lindsay that Queens was part of New York City. We didn't see a snowplow for days, and we had no school for a solid week. Food supplies vanished. Residents were furious, thus the storm's moniker, the "Lindsay storm." As an industrious teen, I sprang into action with my friends. Normally, we got $3 to $5 to shovel a walk and driveway. For this storm, the price was $10 to $15 (come on—it was thirty inches and drifting). Adjusted for inflation, that's about $65 to $100, a ton of money for a fifteen-year-old. My mom made us do elderly neighbors for free, but they ended up tipping nicely. And when the shul's (synagogue's) custodian and snow shoveling service didn't show up, we did that one for free, too. She said Hashem (The Almighty) would reward us for the mitzvah. And indeed, he has, many times over. After a while, a city plow showed up, but it got stuck in the snow. They dispatched another plow to pull it out of the drifts, but the towing chain broke and then that plow got stuck, too. We got to know the drivers, Eddie and Bill. People let them stay warm in their homes, use the facilities and eat a meal.

But after several days, the roads were still impassable, and the grocery stores started running out of food. My parents dispatched my brother, me and some friends to hike to the Bohack's supermarket on Northern Boulevard (about a mile away), where, rumor had it, a supply truck had made it through. Today, word would be spread instantly by social media and iPhone apps; back then, it spread through phone trees.

We loaded up on milk, bread, eggs and cereal and trudged back with our mother lode. We heard stories about price gouging by enterprising teens, but we just charged face value to our neighbors. But folks were really grateful, and we got crazy tips, as much as $5 (over $30 today)! Speaking of enterprising teens, we heard on the radio that motorists were stranded on the Long Island Expressway, a few blocks from our house. We also heard about price gouging: $1 for a plain bagel ($6.50 with inflation), which, at the time, cost pennies.

This really annoyed me, so off I went with friends to the bagel store, which was one of the few places open. We bought a ton of bagels, the owner threw in free cream cheese, and off we went to the LIE, where we gave out the bagels for free. But again, folks were grateful, and we got tips galore, even though we tried not to take them. My mom insisted we put part of the tip money in the tzedakah (charity) box. We also got a bunch of requests for coffee, so for hours, we went back and forth delivering hot coffee. Lessons learned.

Eventually, spring came, and the snow melted. The life lessons learned, however, have endured. First and foremost, you have to be careful what you wish for. Second, it's not always about money; you just have to do what's right and let the chips fall where they may. Third, listen to your mom. She knows what she's talking about.

Dr. Frederick Rotger

Dr. Frederick Rotger is a clinical psychologist, having worked with people suffering from OCD, anxiety, trauma and insomnia for over forty years. Using CBT and ACT wave therapy, he has helped people improve their quality of life by focusing on the present, not the past, and building on their values and inner strengths. His practice is located in Iselin, New Jersey, and he is a resident of Manhattan, New York City. He is a graduate of Rutgers University, and that is where his story begins, on February 9, 1969.

I was a senior at Rutgers College and living in an apartment on Bartlett Street, about two blocks from the College Avenue campus. The apartment was within walking distance of everything on campus, but once the snow had finished falling, it is my recollection that Rutgers was so snowed in that the university canceled classes due to weather for the first time in decades as a result. I don't recall having to clear the snow, which was up to my knees, as our landlord had that responsibility (thankfully).

I do remember watching the news from New York City and seeing the reports of the outcry from the outer boroughs, where secondary streets were impassable and it appeared as though they wouldn't be cleared for days. I remember videos on the local stations of Mayor Lindsay touring the snowed-in neighborhoods and apologizing for the city's failure to clear the streets. Clearly, in those days, with forecasting far less sophisticated than it is now, the city wasn't able to be as prepared for the storm as is possible today. I do recall that it took some time for the streets around Rutgers to be fully cleared as well.

The television forecasts said no snow—until it was up to your knees!

Meteorologist Joe Rao

Joe Rao is one of the most recognized broadcast meteorologists in the New York metropolitan area. Joe was the chief meteorologist and science editor for News Channel 12 in Westchester, New York, for twenty-one years after working for seventeen years as a private forecaster, providing forecasts for over two hundred radio stations across the eastern United States and Canada. He also provided customized forecasts for LaGuardia and Newark Liberty Airports as well as game day outlooks for both the Mets and the Yankees.

An amateur astronomer for over fifty years, Rao joined the staff of the Hayden Planetarium in 1986 as an associate and guest lecturer, a position he holds to this day. Joe has contributed articles about astronomy for Space.com, an online news service, as well as *Natural History* magazine, *Sky and Telescope* and the *Old Farmer's Almanac*. In 2009, Rao was awarded the Popular Writing Award of the Solar Physics Division of the American Astronomical Society for his feature article in the 2008 *Natural History* publication, "Shades of Glory."

In 2023, the International Astronomical Union (IAU) named Asteroid 20009 "Joerao" in his honor.

During my years in junior high school and high school, I would write my weather forecasts out on the blackboard of my home room. It became very popular—so much so that it was referenced not only by other students who used that classroom during the school day but even teachers, who would occasionally stick their heads into that room to see what the weather looked like (especially if we were nearing a weekend).

On the Friday before the infamous snowstorm, I decided to do something a little different. Instead of using the regular white chalk, I used the fat pieces that were in pastel colors. In bright shades of yellow and pink, I wrote—in large letters—"Possible Snow Day on Monday." I then added that a coastal storm on Sunday might deliver enough snow to cause a cancellation of classes on Monday.

Well, this piece of news rapidly spread all over the school like wildfire. By afternoon, people were saying: "Did you hear? Rao says no school on Monday, because of snow."

I particularly remember running into one of the teachers in the hallway, Mr. Richard Russo, who was rather tall and imposing with a deep voice. "I hear you're saying we're going to get a lot of snow this weekend." I nodded a bit sheepishly, to which Mr. Russo then added, "Well, I watched Dr. Frank Field last night, and he said we're going to get rain. Who should I believe?"

All I could say to him was that weather forecasting was very subjective and that what looked to be one thing to one person, could look quite different to another.

Nonetheless, when Sunday came, I woke up and looked out my bedroom window and saw that it was indeed snowing quite steadily. Later, as other members of my family awoke and headed into the kitchen for breakfast, the radio was on, and we kept hearing an announcer continuing to assure listeners that according to the Weather Bureau, the snow would soon change to rain.

But by midday, with more than several inches on the ground and the snow falling even harder, it was becoming increasingly apparent that we were headed for a very heavy amount of snow. By later in the afternoon, we had surpassed the big snow that fell on Christmas Eve night in 1966, and it was looking more and more like we could see more snow even than a twelve-inch blizzard that hit our area two years earlier.

To make a long story short, the final snowfall was in the neighborhood of fifteen to twenty inches. Because everybody believed the snow would change to rain, the Sanitation Department—whose job it was to plow the streets—fell far behind in their clean-up efforts. The city was paralyzed for days, and New York City Mayor John Lindsay took the blame for not staying on top of the unexpected storm. To this day, February 9, 1969, is still known at the "Lindsay storm."

As for me, when the schools finally reopened, I became something of a hero. That's when everybody started telling me that I should become a meteorologist, even Mr. Russo, who said, "You should be on TV doing the weather." I said, "Well, maybe after I go to college and get my degree." And that's when he cut me off and said: "The heck with college—you should be on TV right now! You certainly are better than any of the people on the air right now."

The truth of the matter was, however, that I was a big fan of snow when I was a teenager, and it was quite likely that I was forecasting more with my heart *than with my* head.

A postscript to this story: nine years later, I got a job working for a private weather forecasting service based in Queens. I once told my boss, whose name was Todd Gross, that I thought my forecast accuracy was better than average. Then one day, he came over and dropped a couple of weather maps on my desk, which showed a developing coastal storm. "If you had to make a quick forecast for New York City, what would you call for?" he asked.

I studied them for a few minutes, then said: "I'd initially call for snow, but the area of high pressure to the north of the storm will likely bring in a mild east wind from off the ocean, so probably the snow will change to rain."

Todd let out a guffaw and immediately took back the maps. "Congratulations," he said. "You just forecast what everybody else did for this event." "And what event was that?" I asked.

He replied, "The Lindsay storm!"

Dave Ryan

Dave Ryan, now sixty-eight, and his wife, Judy, are the proprietors of International Physical Therapy in Howell, New Jersey. Dave grew up in the North End of Elizabeth, New Jersey, living at 931 Cross Avenue. As a boy, he delivered newspapers for Academy Newspaper Service, firing the *Elizabeth Daily Journals* and *Newark Star Ledgers* onto front porches, breaking a window or two along the way. He played Little League Baseball with the Lions in the North Elizabeth Youth League. This was a time when kids went out to play at 9:00 a.m. and didn't return home until 5:00 p.m. for supper—no cellphones, no texting, just be home in time for supper (or else).

That Saturday, I was hanging out with the North End gang at Kellogg's Park in Elizabeth, playing basketball, five on five. Then we started a touch football game. I remember it was cold—not bitter cold—and the sun that was out in the morning was gone by afternoon. No one was talking about a snowstorm, yet by Saturday night, you could hear the wind blowing hard. Little did I know, the following day, I would have just about the most miserable experience a human being could have. I woke up to go to church. It was Sunday morning, and it was snowing like crazy. There must have been six or seven inches on the ground by 9:00 a.m.

Afterward, my father informed me we would have to take two shovels and walk the three-quarters of a mile to our other home, which we now rented out, on Jefferson Avenue. Because of the snow, he didn't want to drive the car. My cousin Danny from Dublin was visiting us at the time and he had never seen snow like this. He thought it was great. However, Dad told me there was work to be done. So, my father and I slogged our way in the driving snow, taking an hour to get to the house to dig out the sidewalks and walkway. It was horrible. It took two hours, and whatever we shoveled was immediately replaced with more snow. So, we kept shoveling. It was midafternoon, and there was already a foot on the ground.

Then came the worst part—the walk back. I had icicles forming on my face, and they were stinging terribly. I was almost crying. My coat and hat were soaked, as were my father's. This was "old school." He was afraid that if he had taken the car, which was buried in snow anyway, we would have gotten stuck. Once we got to North Avenue, my father took me into a tavern. He gave me a dollars' worth of nickels so I could play pinball. He let me have one Coke after another as we warmed up while he had a few drinks. When we left, he told me, "Davie, not a word of this to your mother."

STACY GEISENGER

Stacy Geisenger is a renowned blogger who created the diverse *Stacyknows* a leading lifestyle blogazine in Westchester County, New York. *Stacyknows* makes people feel upbeat about themselves and is a go-to source for all things dining, culture, art, health, beauty and fashion. Stacy was eight years old when the borough of Queens was left in virtual isolation for a week after the surprise Lindsay Snowstorm in February 1969.

It was February 1969, I was only eight years old, but I remember the aftermath of the storm very well. It seemed that eastern Queens, Bayside, was not plowed for at least a week. The entire neighborhood was crippled by the surprise and enormity of the storm. The whole neighborhood was out with snow blowers and shovels. Schools were closed indefinitely—there was no Zoom school in 1969. I spent the days playing school over the phone with my bestie Debbie Weissman and watching Mr. Ed *and* The Beverly Hillbillies.

My parents would always drive to New York City together to go to work. But the car wasn't going anywhere. I saw my mother—for the first and only time—walk to Springfield Boulevard to catch the 27 bus to Flushing to take the subway. It is the only time I remember my mother taking public transportation, let alone walk anywhere. It was the sixties, and we didn't know all the health benefits of walking at the time.

5

THE ALAN KASPER SNOWSTORM

JANUARY 19–20, 1978

Following the Lindsay snowstorm in February 1969, there was a nine-year snow drought in New Jersey and the adjacent New York metropolitan area, punctuated by mild winters with well-below-normal snowfall. The winter of 1972–73 established an all-time record for least snowfall in a winter season, with a meager 1.9 inches of snow recorded at Newark Airport. During this period, there were only two snowfalls in excess of six inches, the largest coming on February 12, 1975, when a fast-moving coastal storm resulted in a seven-inch snowfall at Newark Airport.

In the third week of January 1978, there was no indication that the snow drought would be broken. Forecasters and computer model guidance (still in its infancy) did indicate a strong low-pressure system forecast to move up the Eastern Seaboard, yet it was too close to the coastline to bring heavy snow, except in the northwestern counties of Warren and Sussex. Forecasters called for a mainly rain event for New Jersey, with some snow at the storm's onset.

There was no expectation that the first of two crippling snowstorms was about to bury New Jersey.

Meteorologist Joe Cioffi, whose career spans forty years of forecasting on venues such as WPIX, Channel 11, and WNBC television, as well as News Channel 12, Long Island, and New Jersey 101.9 FM, was an aspiring meteorologist at the time. Alan Kasper, the nationally renowned New York and New Jersey meteorologist, was the lone forecaster who felt the storm system could take a colder and more easterly track, and by late Thursday night, he was forecasting heavy snow for much of New Jersey. This author always recalls this storm as the "Alan Kasper snowstorm."

METEOROLOGIST JOE CIOFFI

Once upon a time—a time when there was no internet, with only two computer models that forecast forty-eight hours in advance and the long-range guidance was just one seventy-two-hour upper air map—there was a snowstorm! January 19–20, 1978, brought the largest snowfall in nine years and was the first of two major storms that winter to impact the Northeast in a big way. The more well-known storm was the February 6–8 storm, which is known as the Blizzard of '78.

The January snowstorm, however, was the first major snowfall for the New York City area since the Lindsay storm of February 1969. Much of the region was in the midst of a snow drought. The prior seven winters saw only two six-inch snowfalls in Central Park. In fact, nearly five years lapsed from March 1969 to January 1975, when not a single storm produced more than four inches [of snow]. *When snow lovers want to complain about the lack of snowfall in the winter, try eight winters in a row of below-average snowfall, from 1969–70 to 1976–77.*

This storm was actually the third snowfall in seven days. The first on January 13 produced a four- to six-inch snowfall and a major, memorable ice storm across Long Island. The second on January 16–17 produced a four-inch snowfall that went to rain. This system led to the collapse of the roof of the Hartford Civic Center. Thankfully, there were no injuries.

What was notable about the January 19–20 storm was that the National Weather Service forecast for New York City and nearby New Jersey was for rain developing, possibly beginning as snow or sleet. Remember, folks, this was in a period of history called "pre-cable," when newscasts were on at 6:00 p.m. and 11:00 p.m., and that's it. There was no public access to information unless you had a weather radio (which I did). What followed was a series of weather forecasts that had to be updated every couple of hours. By 8:00 p.m. on Thursday night, January 19, as the snow continued to fall at an inch an hour, the forecast became one to three inches. An hour or so later, two to four inches. The magical moment for me (nineteen years old at the time) was seeing Alan Kasper on WCBS-TV come on at 11:00 p.m. with the following statement: "Well, there are signs that Atlantic City [still snowing at 11:00 p.m.] *is about to lose their low-level cold air. Using the six-hour rule, snow won't start mixing with rain here, if at all, until 4:00 or 5:00 a.m.!"*

You have to understand that for a budding meteorologist, Alan was da man *and a snow lover! Weather in those days got almost no time at all*

Blizzard conditions buried a home in Wanaque, New Jersey, during the blizzard of February 6–7, 1978. *Courtesy of Christopher Stacheleski, NOAA.gov.*

> *on television newscasts. The funny thing is that Joe Rao, my colleague who I didn't know at the time (but who lived just a few blocks away from me), watched the same show and has the same memory. The next morning arrived with a foot-plus on the ground! The snow drought of one-foot snowstorms had ended. Little did we know then that this snow would be washed away a week later by another major storm that buried the Ohio Valley, and then twelve days later, the "Blizzard of '78" would bury everyone from Delaware to New England.*

JOE LAROSA

Joe LaRosa is a survivor of the Great Atlantic Storm of 1962 and an author of two related books: *Our Perfect Storm* (2010) and *Storm Stories* (2019). Joe is a former educator and principal at the Dennis Township Elementary School in Cape May. His boyhood was spent in Sea Isle City, where his father worked for the Sea Isle City Public Works department. LaRosa has a lifetime of storm stories from the oft-battered barrier island.

> *My first meaningful employment after graduating college in 1976 was working for the Cape May County Department of Public Affairs. The*

department was responsible for all of the printing needs of the county, along with promoting the peninsula as a vacation destination. In the summer and fall, I was tasked with, at times, assisting the printer operators, but [I] *mostly* [delivered] *the printed materials and cases upon cases of new Xerox 2024 DP paper to feed the many copiers for assorted departments that keep the county bureaucracy moving.*

The fall and winter were much better, even for the "low man" of the organization. Prior to Christmas, I would travel throughout the county and collect brochures to help promote area attractions as municipalities wanted to distribute these to potential vacationers. Then after the New Year, I would hit the road and travel the eastern United States and Canada in an effort to promote Cape May County.

The government even had a full-time tourism office in Montreal. My hometown of Sea Isle City was a very small, sparsely populated municipality in Cape May County. In January 1978, I was part of a three-person team working an "Outdoor & Vacation Show" at the Farm Show Arena in Harrisburg, Pennsylvania.

On the last day of the show, a major blizzard hit the area, the '78 blizzard, and most of the city was totally shut down. Wondering how things were at home, I telephoned my parents from my hotel. What happened then was a complete surprise. The phone was answered by a very familiar voice. It was not either of my parents or my brother.

Although at first, I didn't recognize the voice, I finally came to the realization that the man on the other end of the line was our family physician, Dr. Francis Hauck. In a time not too long ago, general practitioner physicians actually came to your house when you were sick. In the middle of the blizzard, my mother was sick. Braving a crippling snowstorm in our community, Dr. Hauck eventually arrived at my parents' house to assist my mother. Upon recognizing his voice, I immediately assumed that the worst thing possible had happened.

I was stuck in my hotel in Harrisburg, Pennsylvania, and I was in conversation with a physician who was at my parents' home. Although my mom's situation was not immediately life-threatening, Dr. Hauck advised me to check in with my parents, as I was due home the next day. The next day came, and I was snowed in at Harrisburg. The blizzard had intensified.

The only transportation that was running with any efficiency were the trains. Making sure that my coworkers were okay with me leaving, I boarded a train at the Harrisburg Station. When the train to Philadelphia arrived,

I was treated to a comical sight. I was the only living thing on the desolate platform, yet as I entered the train, the conductor stood out in the gale winds and swirling snow and started announcing: "Train to Philadelphia now boarding, with stops in Elizabethtown, Lancaster—all aboard!"

Prior to leaving Harrisburg, I telephoned my fiancée near Philadelphia and asked her to see if the train was still running from Lindenwold to Ocean City, New Jersey. It was. Upon arriving in Philadelphia, the snow was lightening up as I made my way to the PATCO high-speed line to connect to the Ocean City train. I telephoned my future brother-in-law (in a time before cellphones) and asked him to pick me up at the Ocean City train station.

My brother-in-law was a commercial fisherman who owned a very high four-wheel-drive stake truck. He drove from Sea Isle to the Ocean City Train Station, plowing through drifts to pick me up. Upon arriving in Sea Isle City, I was amazed about the snowfall. Although it was not deep in places, some drifts against houses and fences were chest deep.

I found my parents' house reachable, as my brother who worked for the city public works department had commandeered a snowplow and created an area where, if needed, an ambulance could access the house. Luckily, the ambulance was not necessary.

Upon reaching my apartment a few blocks away, I had to deal with a waist-deep snow drift that had formed at my front door. I did not have a shovel.

6

THE BLIZZARD OF '78

FEBRUARY 6–7, 1978

The occurrence of successive major snow events in New Jersey is fairly uncommon. Heavy snowfalls are usually separated by a month or more, even in very active winters with above-normal snowfall. However, there have been occasions when Old Man Winter brings down the hammer twice in short order. In the snowy winter of 1960–61, there were two substantial snowfalls, 21.4 inches at Newark on December 11–12, 1960, and 15.2 inches on February 3–4. From February 8 to February 11, 1994, in a ferocious winter, 20.6 inches were also recorded at Newark Liberty Airport.

In late January and early February 1978, a near decade-long dearth of snowfall in New Jersey came to an abrupt end. An unexpected winter storm on January 19–20 witnessed ten to fourteen inches of snowfall, which buried much of the Garden State. It was a snowfall that was totally unforeseen. Rain or a bit of snow changing to rain was the official National Weather Service forecast for Thursday evening, January 19. A deepening low-pressure area developed slightly offshore, and the cold air that was in place over New Jersey was not scoured out by milder ocean air. Heavy snow fell Thursday night and into Friday morning. People called out from work. There was a gridlock on the Garden State Parkway and Turnpike, and angry Jerseyans were cursing out the forecasters.

Two weeks later, the weather maps were quiet. However, a tempest was brewing. The northeastern United States was under the control of arctic high pressure, with sunshine and cold daytime temperatures moving into the weekend. An unimpressive low-pressure area and associated

Arctic cold front brought light accumulating snows to the southern Great Lakes. Computer guidance, now becoming a reliable source of data for forecasters, began projecting an intense cyclone forming off the South Carolina coastline.

With arctic air in place and warmer air surging northward and integrating into the developing nor'easter, winter storm watches, which should have been blizzard watches, were hoisted from New Jersey northward into southern New England. This compact, destructive winter storm was well forecasted and fell late Sunday into Monday, which gave New Jersey residents ample time to prepare. By Sunday morning, February 5, winter storm warnings were issued for the entire forecast area, as wind-driven heavy snow began its northward progression, arriving in Southern New Jersey by late Sunday afternoon and moving into Central and North New Jersey after midnight.

New Jersey coastal communities, from Cape May to Sandy Hook, were pounded, as the new moon coinciding with the high tide cycle brought immense amounts of water onto the barrier islands and shoreline. Widespread damages occurred at the shore, with sea walls breaking in several beach towns. The boardwalk at Seaside Heights was badly damaged, and the mile-long boardwalk at Lavallette was destroyed. Heavy snow, unrelenting, fell throughout all twenty-one New Jersey counties all day Monday and into the night. The central pressure of the cyclone dropped from 1,016 millibars to 984 millibars, a rapid deepening widely referred to today as bombogenesis.

This was an extreme blizzard, and it can be characterized only as catastrophic. Seventeen New Jerseyans lost their lives, and there was an estimated $520 million in property damages from Washington, D.C. northward to New England; 20.1 inches fell at Atlantic City, 18 inches in Newark, 22 inches in Sandyston, Sussex County, in far-northwestern New Jersey and 22 inches was recorded at the capital in Trenton. Sheets of snow drifted to massive ten-foot heights, and wind gusts along the coastline were near hurricane force. Homes were destroyed. Newark International Airport closed at noon. Satellite imagery depicted a structure clearly resembling a hurricane-like eyewall.

Hundreds of motorists were immobilized on the New Jersey Turnpike and Garden State Parkway. On Interstate 295 in West Deptford, a tanker jackknifed, spilling seven thousand gallons of heating oil, and resulted in traffic backing up for miles. More than four hundred Ocean and Monmouth County residents were evacuated from their homes due to flooding, and New Jersey Shore communities, like Bradley Beach and Ocean Grove, suffered heavy damage. Overall, the storm damage was estimated to cost $40 million.

Practically all state and local municipalities were closed, with the exception of road crews. Rutgers University canceled classes. Businesses were shut down, and schools were closed. Governor Brendan T. Bryne, two weeks into his second term, worked alone in his office. He had brought a sandwich for lunch.

In terms of sheer fury, this storm was one of the most severe and disruptive ever experienced in New Jersey.

Jack Osborn

Jack Osborn and his father and grandfather were integral parts of the history of the cottage sand beach community that became known as Camp Osborn. He is a Vietnam veteran, having served in the army in an assault helicopter company situated at Bien Hoa in southern Vietnam, not far from Saigon. He was exposed to Agent Orange, resulting in the onset of Parkinson's disease years later.

> *My father was John Osborn Sr. I was named after him, yet we were both called Jack. Our families and many others used to camp out in tents on the beach near the Manasquan Inlet in Point Pleasant until we were told we couldn't use the beach because of planned development. This was in the late 1930s. Many of these families relocated to a sand beach community in Brick on Route 35.*
>
> *Connie and Pella were the original lessees of Cummins and Marion Streets during the late 1940s and 1950s. There were numerous bad storms there. I was in college at the University of Delaware during the March 1962 nor'easter. There was a lot of damage in the Camp; one house was washed out on Elder Street. My mother and father also had a home on the west side, the bay side, of Route 35, which was undamaged in the '62 storm.*
>
> *Next to Camp Osborn, as it became called, was a big house. "The doctor's house." He sold the house to a developer and that is where the Thunderbird Hotel, which is now the Ocean Club, was built, right alongside the Camp.*
>
> *1978 was a bad winter; there was the blizzard in early February. Most of the time it snows at the beach, it turns to rain. Not this time. It began on Monday; it was a terrible storm. There were only a few year-round residents in the Camp who spent winters there, and they had to be dug out. It took days. I have photos of that storm—snowing so hard, sideways, lasting into Tuesday morning.*

There was a storm surge that sent water over the dunes and into the Camp. The dunes always protected the houses in the Camp. If the dunes were breached, it had to be a bad storm. The bungalows were buried in snow, drifting to the rooftops in some houses. Route 35 was impassable; the few travelers who ventured out ended up stranded. Only the '96 blizzard was worse.

Al Mugno

Al Mugno has followed winter storms since he was a boy. He has been a teacher for over thirty years at North Highland Regional High School in North Jersey, and over the years, he has shared his knowledge and passion for meteorology with his students. Growing up on Garret Mountain in West Paterson, he was nine years old during the historic winter of 1977–78, a winter of extreme cold and an unforgettable February blizzard.

I remember the days leading up to this storm, watching every news station weather report, especially my weatherman favorite, Joe Witte on Channel 4. It started out as light snow in the forecasts, which went to a snowstorm two nights before the blizzard. As a young boy, I was hoping for a memorable storm like the blizzard we just had last month, when sixteen inches fell on the mountain. That is what this storm evolved into.

Again, I remember it started snowing on Sunday by late afternoon and that night school was canceled. They were calling for three to six inches. When I got up the next day, my mother and father said, "Your wish has come true—it is a full-on blizzard now forecasted!" That afternoon and night the winds picked up, as did the snow in its intensity. I remember it snowed for a long duration of time, for two days.

Joe Witte showed the satellite imagery. The storm had a formation with a swirl, and I distinctly remember this as it moved in a southwest direction that had TV meteorologists a bit baffled. This is when I found the NOAA radio station on AM to listen to. Our school was closed for three straight days. It took us two days to shovel out our driveway. The snow was up to my mid-thigh and some drifts were over my head. I was only nine and about four feet something, but I can remember a drift that covered our six-foot retaining wall.

The town plows could not keep up with the storm, so I remember they put plows on the garbage trucks to plow us out. On the Wednesday of the storm, when it had stopped, my friend and I measured twenty-four inches, and Garret Mountain observatory had about the same measurement. It was the most memorable snowstorm of my young life, but more would come.

7

THE EXTRAORDINARY APRIL BLIZZARD OF 1982

The winter of 1981–82 was a roller coaster. December was mild, with only 1.4 inches of snow recorded for the month, falling on December 14, at Central Park. January 1982 saw two significant snowstorms, with temperatures substantially below normal for the month; 5.8 inches of snow fell on January 13, followed by 3.5 inches the following day. January 16's daytime high temperature was below thirty-two degrees, and in six days, the thermometer did not reach twenty degrees. January 18 saw a low temperature of negative one at Newark and zero degrees in Central Park.

February and March were uneventful, as a pattern change resulted in sustained above-normal temperatures. The two snowiest months of the year were tame, with almost all of the measured precipitation falling as rain. Only 1.1 inches fell between February 1 and March 31. Spring had arrived in New Jersey—or so we thought.

April arrived with southwest winds bringing mid-spring warmth. On Thursday, April 1, temperatures soared into the upper sixties and lower seventies over New Jersey. However, computer forecast data began recognizing an astonishing climatological sequence of events that would combine to produce an April blizzard that will retain a place in New Jersey history as one of the most remarkable weather events to impact the Garden State. It was the snowiest April ever for New Jersey.

An unseasonably cold—even by early spring standards—arctic air mass over Central Canada began to drain midwinter cold toward the eastern third of the United States. Low pressure began developing over the Gulf

Coast states and moved into the Ohio Valley before pushing east toward the Mid-Atlantic coastline. As early as Thursday, computer models foresaw the cold air driving southward into the Mid-Atlantic, spawning a powerful storm system, a nor'easter, with the probability for a significant winter storm in New Jersey.

Initially, the primary concern was an early spring accumulating snowfall. However, by late Saturday, it was becoming clear that a powerful, even life-threatening, cyclone would imperil the Tri-State region. The rapidly intensifying area of low pressure, loaded with Gulf moisture, rapidly drew the arctic air into its developing circulation. As the weekend drew to a close, there was growing confidence that areas from Central New Jersey northward through New York City, Long Island and southern Connecticut would be facing blizzard conditions on Tuesday.

Good Friday and Easter Sunday were less than a week away. The Yankees home opener against the Texas Rangers was scheduled for Tuesday, April 6. On Monday morning, millions began their commute to work and heard that the National Weather Service had issued a blizzard warning for Tuesday, with eight to twelve inches of snow forecast for northern portions of Central New Jersey and North Jersey, with gale warnings also issued for the coastline.

Residents were told to stay home, if possible, as conditions could become life-threatening during the height of the storm Tuesday morning. Monday was serene, with increasing and thickening cloudiness and cold daytime temperatures, in the mid-forties. The temperature at Central Park and Newark was forty-two degrees at 11:00 p.m. Many New Jerseyans went to bed Monday night wondering how it could snow when it was forty-two degrees.

Precipitation rapidly spread northward into New Jersey before dawn Tuesday, in the form of rain in Elizabeth, Newark and Paterson and wet snow farther inland. By 7:00 a.m., the rain had changed to wet snow as the temperature fell to thirty-four degrees. Morning commuters initially faced wet roads and developing slush. From 8:00 a.m. to 3:00 p.m., as the storm underwent bombogenesis (rapid intensification), very heavy snow developed, falling at the rate of one to two inches per hour. By mid-morning, cracks of thunder and vivid flashes of lightning were observed over northeastern New Jersey and into Manhattan. Temperatures fell into the upper twenties as forty-mile-per-hour wind gusts drove the now-fluffy, dry snow into three- and four-foot drifts.

Drivers faced widespread gridlock conditions on the Garden State Parkway and New Jersey Turnpike. The sole silver lining was that the blizzard was fast-moving; it was a twelve-hour tempest for New Jersey. Had the nor'easter slowed down, as they often do, snowfall amounts of twenty inches or more would have materialized.

By 4:00 p.m., the nor'easter was racing into southern New England, leaving history in its wake—9.6 inches fell in Central Park; 8 to 12 inches in Union County, New Jersey; and over 12 inches in northwestern New Jersey. From Trenton southward, snowfall amounts were far smaller, just 2 to 4 inches of accumulation, as the early rain lasted longer and the storm exited faster. On Wednesday morning, the temperature fell to an unheard-of nineteen degrees in Newark, temperatures worthy of a mid-January morning. To add insult to injury, a weaker storm brought wet snow to New Jersey on April 9, Good Friday.

Al Mugno

Being twelve years old and living on Garret Mountain in West Paterson, New Jersey, Al Mugno was looking forward to Easter week, Easter Sunday and the beginning of the 1982 baseball season. The last thing he was expecting was a blizzard. However, after a quiet winter in Northern New Jersey, with well-below-normal snowfall, the month of April and a powerful coastal storm rivaling the worst that winter can deliver staggered Garden State residents with wind-driven heavy snow and bitterly cold temperatures.

> *The April 1982 blizzard caught just about everyone by surprise and shocked all of us—not just the amount of snow and sleet but also how cold it was for the days before and after the storm. When my mother told me two days before that we can expect a blizzard, I was jumping for joy with excitement and anticipation. She told me right after I woke up, "Guess what? We are going to have a blizzard in two days." And she turned on the radio!*
>
> *I was in disbelief as I called my grandmother and aunt to let them know the news, which they already knew because they were up at about 5:00 a.m. every day. She said, "I know, honey, and I am on my way to the store to stock up on food." She was an old Italian grandmother who could cook up a three-course meal with a few ingredients in no time at all.*
>
> *The blizzard hit Tuesday, and it was snowing fiercely. I went to bed late Monday night because school was canceled. The next day, I got up and went outside around midday to help shovel the driveway. The snow*

was about twelve inches deep then. The winds were howling on Garret Mountain, and it was a tough go so I abandoned this and decided to sled on our hill, which was not plowed. As I got my sled out, the snow began to mix with sleet. And then as I was trying to sled, it changed to sleet. A wind-driven sleet blizzard actually started to hurt, so I abandoned this and went inside for some hot chocolate, and my father made a fire in our fireplace. We could hear the sleet hitting the windows and side of our home. No baseball!

Nick Stefano

Nick Stefano is the president of the North Jersey Weather Observers and is chairman of the NJWO Standardization Committee, which oversees the instrumentation used as well as the accuracy of reports submitted during significant weather events. A resident of Sparta in Sussex County, he previously lived in Wantage in far-northwestern New Jersey for thirty years. He has maintained daily weather records for decades and has provided vital storm information through television and social media to municipalities within the county.

Stefano was a college student during the spring of 1982, as confidence grew during the first weekend of April that a historic spring blizzard was a distinct possibility, and by Monday night, the storm was on New Jersey's doorstep.

I was a student at William Paterson University that spring. I played college ball, and the team had returned from our Florida spring training baseball trip a week earlier. We played one game when we returned, and then the storm hit!

A foot of snow fell in the New York metro area and Northern New Jersey after a period of rain at the start. I recorded 12.5 inches in North Haledon (Passaic County), and *the opening-day Yankees–Texas Rangers opening season game at Yankees Stadium was snowed out. Later that day, a Tuesday, the temperature fell to sixteen degrees, an all-time record low for North Haledon; 1.21 inches of melted precipitation fell from the pre-dawn hours until the storm ended by midafternoon.*

8

THE MEGALOPOLITAN BLIZZARD

FEBRUARY 11–12, 1983

New Jersey State Police and National Guard units brought 348 people stranded on the Garden State Parkway to an armory in East Orange, where they spent Friday night. "There were a lot of children, and they were crying,"
—John Hargadon, sixty, Fair Lawn

To be classified as a blizzard, a relatively rare occurrence in New Jersey, a winter storm must produce sustained winds of over thirty-five miles per hour and visibilities under a quarter of a mile, combined with falling snow. By this criterion, you can have a blizzard with only three inches of snow falling. Blizzards along the Eastern Seaboard are life-threatening and destructive societal events, and from a meteorological vantage point, they are very difficult to forecast.

Blizzard conditions can occur over small portions of the overall forecast area or cripple vast geographical areas. The Blizzard of February 10–12, 1983, was historic and unique, as it crippled major metropolitan areas from Virginia to New England. Richmond, Washington, Baltimore, Philadelphia, New York City and Boston felt the fury of this mighty storm, which resulted in forty-six deaths and stranded motorists, immobilized millions in their homes and caused millions of dollars' worth of damages along the Interstate 95 corridor.

That is why this historic storm earned the moniker the "Megalopolitan Blizzard."

The great blizzard dropped 17.8 inches of snow in Linden and 25 inches in Trenton and Philadelphia, with widespread 15- to 20-inch accumulations experienced throughout New Jersey. The storm forced President Reagan and First Lady Nancy Reagan to cancel a planned visit to Camp David. The president remained in the White House and had dinner with Secretary of State George Shultz to discuss easing tensions with the Soviet Union, which led to the Reagan-Gorbachev summits and the end of the Cold War.

The storm was very difficult to forecast, as forecasters had concerns that the very cold air in place over the Northeast could suppress the heavy snow to the south, as computer models suggested, causing uncertainty in snowfall accumulation forecasts for New York City and Northern New Jersey. The massive snow shield inched up the Eastern Seaboard, reaching Southern New Jersey by Friday morning and enveloping the entire state by early afternoon. Snowfall rates of three inches per hour were observed Friday night into early Saturday.

The monster blizzard crippled traffic, with the New Jersey State Police assisting hundreds of stranded motorists on the Turnpike and the Garden State Parkway. On Interstate 295, a tanker collided with a tractor trailer on Interstate 295, spilling five thousand gallons of heating oil and backing up traffic for miles on the snow-covered road.

County and state roads were impassable in the face of blinding snow and wind. Thundersnow late Friday night saw vivid bolts of lightning illuminate the skies over Union and Essex Counties as an intense snow band set up over northeastern New Jersey into New York City.

It was truly a historic and crippling blizzard.

METEOROLOGIST JOE RAO

It's absolutely unbelievable, folks. Just watching this area of snow blossoming south and west of New York City. Twelve inches on the ground in Baltimore, ten inches in Washington. Some Washington suburbs may have twenty-four inches of snow on the ground tomorrow morning. The storm literally is going out of control now, moving off of the Mid-Atlantic shoreline. Snow that started here about an hour ago in New York City will fall, moderate to heavy, for at least the next eighteen to twenty-four hours. We don't see a break in this until one or two o'clock Saturday afternoon. Sunshine will return on Sunday, which will be a dig-out day.

—WABC 2:30 p.m. audio

Meteorologist Joe Rao was on the air on WABC 770 News Friday afternoon, February 11, 1983, as the Megalopolitan Blizzard enveloped New Jersey and New York City. He shared the microphone with Dr. Judith Kuriansky and Alan B. Colmes throughout Friday night, providing hour-by-hour storm updates. At the 2:30 p.m. news break, Rao provided an up-to-the minute blizzard update (see page 90).

> *A portion of the 5:00 a.m. weather discussion from the National Weather Service in New York City* [which, at that time, was located in Rockefeller Center]. *Note in the second paragraph that the expectation* [based on the computer guidance forecasts that came out several hours earlier] *that the storm will not come quite as close to us as earlier expected, and we will be on the northern fringe of the storm system's heavy snow shield.*
>
> *By mid-morning, the National Weather Discussion from the Hydrometeorological Prediction Center in Washington, D.C., called attention to the historic nature of the storm and, like everybody else, was thinking that New England would be spared (Boston ultimately got over a foot of snow). It's interesting to page through all the reports and summations and see how all the weather offices were backpedaling and revising their original forecasts by midday and early afternoon, when it was becoming increasingly obvious that the storm's effects would be felt much farther north than originally anticipated.*
>
> *I don't know if I'll be around to see my great-grandchildren, but I just became a grandfather several months ago, and you can bet when he gets old enough, I will sit him down and tell Julian the story of how "Pop-Pop" covered this amazing snowstorm.*
>
> *Apparently, after the storm had ended, the New York City and North Jersey tabloids were all over the NYC Weather Service Office for completely blowing the storm prediction. This led to a long statement by Tom Grant, one of their lead forecasters, chronologically detailing all of the forecasts leading up and continuing through the storm. When finished, Tom (who was passionate about his job) lashed out at the newspapers' "sloppy reporting." Incidentally, Tom was also an adjunct professor at my college (CCNY). We got to be very good friends, although I lost track of him after the NWS moved from New York City to Eastern Long Island in the '90s. I think that's when Tom retired (he lived in New Hyde Park). I don't know if he's still alive. If he is, he has to be somewhere in his nineties.*

By midnight, I had upped my final accumulation estimate to eighteen to twenty-four inches. In addition to radio, I was also keeping the NYC Sanitation Department constantly assessed of the situation using a TWX teletype machine. Rather than use numbers, we listed each winter storm in 1982–83 with names; we named the Megalopolitan Blizzard "Earl."

I did my final broadcast at around 5:00 a.m. Saturday and collapsed on the office couch. When I woke up at around 12:30 p.m., the snow was over and the sun was out. I went out to dig my car, a 1973 Dodge Polara, a big car, out but found it hopelessly covered by a huge snow drift. A guy in a pick-up truck offered to pull my car out for twenty dollars. He attached a chain to the front of the car and pulled it out in an explosion of puffy and powdery snow. I then got in and negotiated the local streets in Flushing to the Whitestone Expressway, over the Whitestone Bridge and finally home in Throgs Neck.

Quite an experience.

Andrea Sabin Barrall

Andrea Sabin Barrall is the Director of Marketing for Swift Electric Supply, based in Teterboro, New Jersey. She was formerly the marketing director for Turtle Inc. for thirty-one years. An industry-recognized marketing specialist with a background in corporate event planning, networking and team building, she is symbolic of New Jersey businesswomen making their mark at the highest levels of corporate industry. She lives in Edison Township with her husband, Jim, an engineer. They are the parents of two daughters, Ashleigh and Samantha.

In 1983, I was a student at Montclair State College. We were aware there was a big snowstorm coming, yet no one was sure how much. The snow began Friday afternoon and lasted well into Saturday, with approximately eighteen inches of accumulation. The snowdrifts on campus, on top of the Watchung Mountains, were up to our heads. We were stranded on campus with no classes.

Luckily, dining services still operated for campus residents. My roommate and I and other friends had welcomed an exchange student into our group. It was her first snowstorm, and she was so excited. She wanted to "feel" the snow in every way. Without a coat on, she grabbed one of the cafeteria trays and used it as a sled. Her delight made us look at the situation as a joyful opportunity instead of a burden.

KEN PILEGGI

Ken Pileggi was the assistant branch manager for Cooper Electric Supply in Linden, New Jersey, before he became a co-owner of SPT Electric Supply in Bridgewater. He had been following the National Weather Service forecasts closely Thursday and Friday morning, which were wavering on how far the northern extent of the heavy snow would advance. The forecast snowfall would impact company electrical supply customer deliveries that Friday afternoon, as well as counter pick-up business. He was busy calling electrical contractors he knew in nearby Middlesex and Monmouth Counties, asking if the snow had started yet. There was a tension in the air as the great blizzard inched northward through New Jersey.

> *It was overcast when I drove to work on Friday, and it was very cold, in the lower twenties, and quite breezy. The sun was still barely visible through the clouds. By mid-morning, there was a scattering of snow flurries, on and off, yet no sign of steady snow. The counter was busy with a lot of contractors trying to get ahead of the storm, yet no one was sure how much snow we would get. The forecasters were going back and forth, some saying the storm track may stay south and New York and parts of Jersey may be spared.*
>
> *I had my head counterman begin making phone calls to nearby contractors to see if the snow had started. I heard on WCBS that there was heavy snow in Baltimore and Philadelphia. I did the same, I called Iselin Electric, and the secretary told me it was starting to snow hard. Then I called Vince Orazi of Orazi Electric, whose shop was in nearby Colonia. Vince was in his office, working on estimates. It was a conversation I remember to this day. Vince, in his gruff voice, said, "Snow—we have two inches on the ground, and it's snowing sideways." It was snowing so hard he was going to close up shop, yet there was still barely a snowflake in Linden.*
>
> *Colonia was only five miles away from Cooper. I had taken my lunch, and by noon, the snow flurries were more numerous. By now, customers were coming into the counter telling us about whiteout conditions on the Parkway with snow piling up at the shore. Then, just a few minutes before 1:00 p.m., the "wall" arrived.*
>
> *It was literally a wall of snow. In a minute, the flurries became steady, then heavy snow. It seemed like we had an inch in fifteen minutes. By midafternoon, it was snowing more than two inches an hour.*
>
> *We stayed open; that's how it is in the electrical distribution business. There were still a few contractors coming in. By 5:00 p.m., there was ten*

> *inches of snow on the ground, and I had to dig my car out of a nearly three-foot drift.*
>
> *My fiancée, Eileen, and I were dating at the time. I felt the need to visit her that night. My mother was not happy with me. No one was on the road except snowplows and idiots. I didn't have a plow. It snowed hard into Saturday morning; nearly twenty inches were on the ground, yet it was hard to tell with the big drifts.*

Al Mugno

Al Mugno was an eighth-grade student in early February 1983, and the television and radio meteorologists were filling the airwaves with talk of a major East Coast snowstorm, a blizzard. Washington, Baltimore, Philadelphia and South Jersey were going to be buried, yet the forecasters did not believe the massive shield of heavy snow would reach Northern New Jersey and New York City. Very cold, arctic air was entrenched over the Northeast and seemed likely to block the northward advancement of the shield of heavy snow.

> *I was in the eighth grade and could remember two days before the Friday storm. They said it was a possible snowstorm with four to six inches expected. A winter storm watch was put up the day before, on Thursday. The excitement that morning in school was high—a three-day weekend! When the news came on, it was like a dagger; the winter storm watch was canceled, and we would see maybe one to three inches of snow. What a letdown! Friday morning, as we ate breakfast, we heard again on CBS radio that it would be a one- to three-inch snowstorm, and a travel advisory was issued.*
>
> *The storm was going out to sea and would just brush us. At lunch, I asked my teacher what the latest was on the storm. I was hoping it would turn north and not go out to sea and miss us. To my surprise, he said, "I called my friend in Trenton, and it was moderately to heavy wind-driven snow." Mr. Lijoi made my day when he said it should be starting here shortly and that the forecast had changed to a snowstorm!*
>
> *At about 2:00 p.m., the flakes started flying, and by 2:30 p.m., the ground was completely covered. It started snowing heavy, with windy conditions. We were called from our classes to go back to homeroom, and* [we were told] *that parents were instructed to pick us up. My teacher*

Mr. Lijoi heard on the radio there was a blizzard warning that they were expecting up to twelve to twenty inches of snow!

Our ride came at 3:00 p.m., and there was about two inches of snow on the ground with blizzard-like conditions setting in. When we got home, we opened the shades to our sliding glass door and watched the blizzard unfold all night. When I was going to bed, I turned the TV on in our room, and the 10:00 p.m. news came on, Channel 5, and there was a report of six inches of snow in an hour in Philadelphia! We had about three to four inches in that same hour, from 8:00 to 9:00 p.m.

It was an amazing storm that just dumped snow in a short time span, and we wound up with about twenty inches [of snow]. *The snowplows got stuck on our hill, and they had to get the big international trucks to plow our streets. The next day, we went outside, and the snow was over my knees.*

9

THE THANKSGIVING SNOWSTORM OF 1989

The Currier and Ives *Home to Thanksgiving* portrayal of a snow-covered countryside home has long evoked hopes for a white Thanksgiving and the beginning of the holiday season. In New Jersey, measurable snow on Thanksgiving Day is an extremely rare occurrence, having occurred only four times since 1898. The largest Turkey Day snowfall, 3.9 inches, occurred in 1938.

As Thanksgiving Day 1989 neared, a more midwinter weather pattern became established over New Jersey. Unseasonably cold air was in place, being channeled into the Northeast by a cold Canadian high-pressure system over southern Canada. A potent early winter storm developed in the Tennessee valley and moved northeastward to the Virginia coastline; it was a classic winter storm along the snow-friendly 40–70 benchmark.

Accumulating snow spread northward into the Mid-Atlantic states. Washington, D.C., experienced its first ever white Thanksgiving and first ever significant accumulating snowfall in November. By late Wednesday evening, November 22, moderate snow had developed over Southern New Jersey and advanced steadily northward, reaching Linden and Newark by 10:00 p.m.

WCBS 880's chief meteorologist Craig Allen told listeners during his 5:00 p.m. forecast on the day before Thanksgiving, "Folks, we're going to have a white Thanksgiving." The snow fell steadily, at times heavily, through all of Thanksgiving morning; 1.8 million viewers at the annual Macy's Thanksgiving Day Parade endured blowing snow and temperatures in the mid-twenties with wind chill temperatures in the teens. Although it was not a powerful storm, it established records for early season snowfall from Washington northward into southern New England.

As the snowstorm was getting underway, a major fire destroyed the historic Jenkinson's Pavilion in Point Pleasant Beach. The fire was not directly related to the storm, yet it posed a challenge to the firefighters who battled the Thanksgiving Eve blaze.

Six inches of snow fell in Newark, and widespread three- to five-inch accumulations were observed throughout the Garden State, including from the barrier islands southward to Cape May. It was a statewide snow event, more typical of mid-January than late November, and it was the largest Thanksgiving snowfall on record in New Jersey.

METEOROLOGIST JOE CIOFFI

Veteran meteorologist Joe Cioffi remembers that historic Thanksgiving winter storm, the largest Thanksgiving snow accumulation since weather records have been kept. He was a television forecaster for News Channel 12, Long Island, at the time, when it became clear in the preceding days that New Jersey and the entire Tri-State area would experience a midwinter snowstorm on Thanksgiving Day.

> *The Thanksgiving snowfall of four to eight inches in 1989 was remarkable for a number of reasons. Firstly, it was the first November accumulating snowfall in ten years. Second was that this snowfall wound up being the biggest snowfall of the entire winter. Third was that Friday morning, the day after Thanksgiving, the morning low at my house was eleven* [degrees] *above zero!*
>
> *This turned out to the lowest temperature recorded for the day in Islip (I lived two miles from the airport). It also reached fifty-five* [degrees] *by afternoon, melting most of the eight inches* [of snow] *I received. That eleven-degree temperature was also colder than any morning low temperature for the entire winter season! I normally worked Thanksgiving Day and took the day after off. I worked the morning shift, as I always worked the holiday. Once I was done, I had the rest of the day off for family gatherings. I remember driving into work with snow falling. In the days before, it seemed like this unusually early storm system was indeed going to happen.*

An unusually cold high-pressure system had pushed into Upstate New York and eastern New England. Temperatures were cold for late November and the day before stayed mostly in the thirties. Once the snow started falling Wednesday, as low pressure intensified to the south, just offshore, temperatures fell into the low to mid-twenties. Thanksgiving morning, it was snowing steadily all morning before it finally ended around lunchtime. The rest of the day was cloudy and cold with a north-northeast wind and temperatures in the twenties. It might as well have been January!

I was supposed to drive to my parents' house for Thanksgiving dinner, but due to the slippery roads—and I had a ten-month-old baby—we ran out and bought an emergency turkey and had our own quiet Thanksgiving.

Al Mugno

Now a high school teacher at Northern Highlands Regional High School in Allendale, New Jersey, Al Mugno was in college in the fall of 1989, looking forward to the family Thanksgiving gathering at his home on Garrett Mountain in West Paterson, New Jersey. During the beginning of that Thanksgiving week, an early season winter storm was on the weather maps, and the possibility of the first white Thanksgiving in decades was making headlines.

I was a junior in college and remember, two days before, on the news, [they were] *saying there would be some snow, more in the suburbs and rural regions of Northern New Jersey. We went out the night before Thanksgiving, as most college students do with everyone being home for the holiday. That night when we left the school, we were celebrating as the snow started to fall late Thanksgiving Eve.*

We were like little kids in enjoyment, as most are with the first snowfall of the season. I remember going home; [I] *went to bed and woke up Thanksgiving morning, and to my surprise, we had about four and a half inches at West Paterson. All the high school Thanksgiving Day games were canceled, and we had to shovel out to go to our grandmother's home for Thanksgiving. It was a classic Thanksgiving and a great surprise storm to kick off* [the] *holidays and the winter.*

10

THE STORM OF THE CENTURY

One forecaster stated, "This is the storm, not of the century, of the history of mankind.... The heaviest snowfall was at Newfound Gap, where US Highway 441 crosses the Tennessee and North Carolina border, recorded five feet of snow and drifts up to 14 feet were observed at Mount Mitchell. There were wind gusts there as high as 110 mph according to NDC."

When forecasters and broadcast media began referring to this massive, destructive storm system as the "Storm of the Century" in early March 1993, it was not the usual pre-snowstorm hysteria and misinformation that invades the airwaves and often competes with official forecast information.

Its only competitors for New Jersey's worst storm ever were the Blizzard of 1888 and the Ash Wednesday storm, both of which occurred in March and had far less geographical impact. The March 12–14, 1993 cyclone carved a deadly path across twenty-two states and impacted 40 percent of the United States' population, 120 million people.

Birmingham, Alabama, recorded thirteen inches of snowfall, and accumulating snow was reported in all sixty-seven Alabama counties. Up to half a foot fell in parts of the western Florida panhandle. Fifty-six inches were measured on Mount LeConte in Tennessee, and fifty inches fell in Mount Mitchell, North Carolina, with fourteen-foot drifts recorded.

Near-hurricane conditions occurred over Florida, as the explosively intensifying area of low pressure moved across Florida and turned north, up and parallel to the Eastern Seaboard. Dry Tortugas, west of Key West, Florida, recorded a wind gust of 109 miles per hour. Taylor County, Florida, witnessed a twelve-foot storm surge, and the superstorm spawned over a dozen tornadoes in Florida, resulting in forty-four deaths. The storm had a central barometric pressure equal to that of a category 3 hurricane. Wind gusts of over 70 miles per hour were observed at fifteen recording stations along the Eastern Seaboard. All the major airports along the Eastern Seaboard into Halifax, Canada, were closed, with thousands of flight delays and cancelations. According to NOAA:

> *An estimated $5.5 billion dollars in damages occurred from Florida to Maine with 10 million people losing power, including 2 million residents of New Jersey, an unprecedented and destructive legacy. In excess of 270 people perished during the storm, including 48 people who died in the Gulf of Mexico, Atlantic Ocean and Canadian Maritimes. Many deaths resulted from heart attacks shoveling snow. New Jersey Governor Jim Florio declared a state of emergency at 11:00 a.m. on Saturday, March 13, asking residents to stay off the roads and highways. Florio said, "It is the worst snowstorm in the history of New Jersey," which, in terms of snow accumulation, wasn't accurate. Incredibly, all interstates north of Atlanta, Georgia, were closed.*

By Thursday evening, a winter storm watch had been posted for all of New Jersey, with a blizzard watch in effect for much of Central and North New Jersey. By the time the 5:00 p.m. forecast was issued for the following day, Friday, March 12, winter storm warnings, blizzard warnings and gale warnings were issued for all twenty-one New Jersey counties, a forewarning for tens of millions that perhaps the most destructive winter storm in the past one hundred years was less than twenty-four hours away.

New Jersey was roundly battered by the storm, which edged just off the New Jersey shoreline as it moved north-northeastward. Heavy snow arrived from the south and moved north early Saturday, arriving first in Cape May and accumulating up to four inches in coastal locations before changing to sleet and then heavy rain driven by sustained winds of over forty miles per hour. Ten to twelve inches accumulated in Central New Jersey, with fifteen to twenty inches falling over northern and northwestern New Jersey. The snow changed to a stinging sleet by Saturday afternoon in most New Jersey locales and then changed back to snow before finally ending early Sunday morning.

There may never be another winter storm like it!

A person in the distance navigates a snow-covered rural road as the "storm of the century" brings record snowfall, damages and flooding to the eastern third of the United States. *Courtesy of Christopher Stacheleski, NOAA.gov.*

Meteorologist Joe Rao

Meteorologist Joe Rao was working on Friday, March 12, at Long Island News Channel 12, as the powerful storm bore down on New York and New Jersey. Blizzard warnings and high wind warnings had been issued Friday evening by the National Weather Service in advance of this winter storm of historic proportions. Social media was in its infancy in 1993, so the high-voltage hype machine that preceded winter storms in the 2000s was not present.

> *I called for blizzard conditions and eight to fourteen inches of snow with my first Saturday forecast, with a changeover to ice and plain rain for most areas by afternoon. Indeed, by 3:00 p.m., one of our reporters, Nicole Nogood, came into the studio, having just spent the earlier part of the day outside, and said that the sleet was coming down very hard, "like machine gun bullets," and she had to head indoors because it was painful to be outside and be subjected to the ice hitting her.*

The sleet compacted the earlier snow that had fallen and cut back on the snow accumulations—ten inches had already fallen. I remember spending most of the next day with a shovel, chipping and scraping away the heavy, wet snow in my driveway that had fallen the previous day. It had a consistency similar to concrete. Quite a storm.

METEOROLOGIST CRAIG ALLEN

Craig Allen, longtime chief meteorologist on WCBS 880 and CBS television, had forewarned listeners as early as Monday that there were signals that a major East Coast winter storm would be heading toward the Tri-State area by the weekend. Model guidance had indicated, consistently, up to five days in advance of the storm, that the entire Eastern Seaboard, from Florida to Maine, would be imperiled by a dangerous winter storm of historic proportions.

It was controlled chaos; they knew they needed a third body, so I was in to help with the coverage. I mostly did radio from that weather office; I was on the other side of the weather office for WCBS 880. Dr. Frank Field was in the studio, and his son, fellow meteorologist Storm Field, was on the other side of the weather office in front of an old monitor.

Anytime I was free, I came over and leaned into the shot while Storm held the microphone, and we provided, as best we could, up-to-the-minute storm information. The biggest question we all had to answer was how fast and exactly where the heavy snow would change to sleet. The change to sleet and rain had already taken place at the Jersey Shore and south-facing shores on Long Island. I remember the changeover occurred pretty quickly for the coast, with heavy snow lasting longer for the city and Northern New Jersey.

STEVE DEDINSKY

Lifelong Linden resident Steve Dedinsky and his wife, Donna, then newlyweds, had just moved to their new home on Rosewood Terrace in Linden, New Jersey, in 1991. Donna was employed by Wakefern Corporation, which owns multiple Shoprite supermarkets in the area. They were making storm preparations after hearing dire storm forecasts on television and radio. Steve worked as a medical electronics technician, repairing X-ray machines for Philip Inc.

> *We had only been married for a few years and had just bought the house. All week long, we kept hearing about this big storm for the weekend—that it could be life threatening, maybe the worst ever. We finally went to ShopRite Friday night to get some food in the house for the storm. That was a big mistake; the line of shoppers extended all the way to the back of the store and then snaked around in a giant circle.*
>
> *There must have been fifty or sixty shoppers ahead of us, with carts filled to the brim. It took two and a half hours to finally check out and get home. We listened to the late-night news and the forecast only got worse—over a foot of snow coming, followed by ice and sixty-miles-per-hour winds.*
>
> *We went to bed, and by early morning, the snow had started. I got up at 8:00 a.m., and there was already at least six inches on the ground. The winds were adhering the snow to the utility poles, and it was already piling up in drifts. By noon, nearly a foot had fallen, and I went outside, trying to clear the sidewalk and driveway. It was a waste of time. I could barely see the house it was snowing so hard—and snowing sideways.*
>
> *It was no use. I went inside for lunch, and when I went back out, the snow had changed to heavy ice, sleet. It was brutal to endure, stinging my face to the point I gave up. The storm got worse, and then sleet piled a sheet of ice, about two inches, on top of the snow. Then the temperature began dropping; by night, it was in the teens.*
>
> *I didn't have a snowblower; whenever I needed snow moved, my neighbor would lend me his snowblower or clear our walkway and driveway for us. The next day was Sunday, and everything was shrouded in ice; it was glacial. I spent hours chipping and chopping away, working on our cars so we could get into work Monday. One of the pipes had burst; there was water in the basement. I also had to check on my mom, who lived alone on Gibbons Street, and had to help her get her sidewalks cleared. It was fourteen degrees Sunday morning, with bright sunshine.*
>
> *I've seen storms with more snow, yet this storm was an assault.*

Al Mugno

Teaching in Jersey City in 1993, Al Mugno was himself a student of meteorology and had chronicled the most brutal winter storms that had ever laid siege on New Jersey. During the second week of March 1993, New Jerseyans heard the sound of distant cannons as it became clear that a classic combination of arctic cold, moisture-laden air from the Deep South and the northern and southern jet streams were converging to produce a winter storm for the ages.

> *I was teaching in Jersey City at the time and was an avid Weather Channel watcher. I can remember in the five-day forecast showing snow for Friday, March 12. Then they had on their winter weather expert, who said this could be an intense storm if all three jet streams converge over the Deep South. This really fascinated me because I never knew of the three jet structures.*
>
> *Every morning before work leading up to the weekend, I had the Weather Channel on. It showed snow and a mix. On the Weather Channel, as early as Wednesday morning, they were saying that it was shaping up to be a superstorm, as their winter weather expert had discussed previously. I remember them showing how and where the jets would converge to make this a dangerous storm from Florida through New England, and* [they said] *we could expect blizzard-like conditions from Northern New Jersey through Maine.*
>
> *The storm surge was going to be an underestimated phenomenon; it tore up parts of the Jersey Shore. I remember watching Friday night with our floodlights on, watching the blizzard unfold and the winds howl. Midway through Saturday morning, the snow changed to sleet, and it was sleeting in buckets. You could hear it clanking off windows and the siding of the home.*
>
> *When Sunday came along and we started to dig out, I measured fourteen inches of snow and sleet. The snow was so heavy, it was difficult to shovel and move. I remember getting the phone call that school was canceled Monday. Then on Monday, after digging out more and finally making it to the roadway from our driveway, we were called again, and school was canceled Tuesday as well. The sleet was impressive with this storm; it was actually painful, as were the relentless winds, over sixty miles per hour.*

Dave Ryan

Spring was days away, and Dave Ryan's beloved St. Louis Cardinals were in the thick of spring training, as the 1993 baseball season loomed. Dave; his wife, Judy, a licensed physical therapist; and their nine-year-old son, Danny, were preparing to move from their home in Cranbury to a townhouse in Freehold when their plans to visit the property on the second Saturday in March were abruptly interrupted by a powerful nor'easter that was moving up the Eastern Seaboard, one of the most powerful winter storms to ever impact New Jersey.

> *If you're ever in the mood to just sit home all day and stare out the window, you may want to witness a daytime snowstorm in Jersey. I was living in Cranbury, New Jersey with my wife and nine-year-old son, Danny, in March 1993. We had purchased a new town home in Freehold just the month before. The home was still under construction, so we had made plans to visit the site that Saturday. The forecast changed all that in an instant. A blizzard warning threw the populace into full-fledged hysteria. Instead, with blinding snow falling and five inches already on the ground, I was at the local Shop Rite at 7:00 a.m., panic buying with what seemed like 90 percent of Cranbury. The storm was a quick-mover and was over by about 7:00 p.m. that night, but what a storm it was! It dumped at least twelve inches of snow, followed by hours of sleet, and was accompanied by howling winds. The snow drifted to at least three feet and then was hardened by a coat of ice.*
>
> *At the conclusion of Mother Nature's scolding, I decided to walk into town to pick up Chinese food, which happened to be the only establishment open. But plummeting temps after the storm totally froze the top layer of snow, forming a crust that made walking perilous. The next day, I, along with the other tenants of the condo complex, spent hours trying to clear off cars that seemed to be encased in a frozen tomb. The temperature was in the mid-teens with wind chills below zero.*

11
THE BLIZZARD OF '96

"There's no place to eat, no place to sit—it's ridiculous," said stranded Renee Harrielal. A train announcer, Mike Alfano, counseled patience: "It doesn't look like it'll get better any time soon. There's not much you can do besides wait." The region had been ready for the storm, but it didn't matter. The blizzard continued overnight and through Monday, thirty-seven hours in all, which kept most plowing equipment off even major highways until the next day. Snow fell at a rate of two inches an hour, and anyone who scraped off the car before Monday night ended up having to do it again.

The Blizzard of January 7–8, 1996, in terms of its impact on the lives of New Jerseyans—the loss of life, the widespread destruction and property loss—was one of three snowstorms to be ranked "extreme" on the Northeast Snowfall Impact Scale (NESIS). Over 150 lives were lost along the Eastern Seaboard, including 30 from associated flooding. And $3 billion in property damage was reported along coastal and inland areas. Thousands of citizens lost power, with outages reported in all twenty-one New Jersey counties.

It was a "top five" blizzard in New Jersey history, and it caught no one by surprise. Forecast models, still an evolving forecasting tool, had locked onto the idea of a powerful East Coast snowstorm, a blizzard, as early as New Year's Day. As the week progressed, the threat grew. Local and state officials began coordination efforts to prepare for the likelihood of twenty to thirty inches of snowfall, or more, from Cape May to High Point.

The message was clear: this would be a life-threatening storm.

The climatological scenario was classic for storm development along the Eastern Seaboard. Arctic high pressure funneled very cold air southward into the Northeast and Mid-Atlantic regions. Low pressure advanced eastward from the western Gulf of Mexico to the Georgia coastline and began edging, like a tremendous machine, slowly up the Eastern Seaboard. A blizzard watch had been issued Saturday morning, January 6, for New Jersey. New Jersey 101.5 chief meteorologist Alan Kasper's morning forecast reflected the approaching blizzard. A respected meteorologist nationwide, Kasper forecast eighteen to twenty-four inches over South New Jersey, fourteen to eighteen inches over Central and North New Jersey and up to ten inches over the northwestern counties.

There was some concern that the cold, dense arctic air entrenched over New Jersey could suppress the heaviest snowfall south, over Baltimore and Washington, D.C. This was one winter storm in which the usual issues about where the rain-snow line would set up and if milder easterly winds change the snow to rain in coastal areas were not concerns. This was going to be a full-fledged blizzard for the entire state.

By Saturday evening, the blizzard watch was upgraded to a warning. There was no escape. Heavy snow crept into South New Jersey before daybreak Sunday, spreading slowly northward into Linden and Newark by early afternoon. For twenty-four hours, wind-lashed heavy snow pounded the Garden State, with mesoscale snow banding producing thundersnow, with three-inch-per-hour snowfall rates over Central and North New Jersey late Sunday night.

From Cape May County and the barrier islands in extreme southeastern New Jersey to High Point State Park in the northwest tip of the state, the statewide "state of emergency" restricted travel; even a trip to the local supermarket or pharmacy was perilous, as millions of New Jerseyans were homebound. Twenty-eight storm-associated fatalities were reported statewide, many resulting from heart attacks while shoveling snow.

The New Jersey Turnpike was closed to all traffic but emergency service vehicles. The National Guard was deployed to rescue state troopers and assist stranded motorists and medical personnel as the blizzard persisted into Monday afternoon. Drifts up to six feet deep made travel in rural areas impossible. Although thousands of New Jersey, Connecticut and New York City businesses were closed, for many, calling out of work and staying home was not an option, and police, fire and hospital personnel reported for duty in the face of near-impossible travel conditions.

In Newark, 27.6 inches of snowfall was recorded, 22 inches were recorded in Linden and 20.2 inches were recorded in Central Park. An astonishing 30.7 inches were observed at Philadelphia, and 25 inches were recorded across the river in Trenton. Winds gusting to gale force drove the powdery snowfall into road-blocking drifts.

Several hundred commuters were stranded on Monday at a train station in Trenton. The restaurant inside the station was closed, and travelers couldn't even get a cup of coffee. Most schools and government offices were closed for a week, cutting off $1 billion in New Jersey commerce. The National Guard was forced into service to rescue state troopers in police cruisers. Houses were drifted over, and neighborhoods looked like ghost towns.

Children will remember the Blizzard of '96 the way remembrances of the Blizzard of 1888 were passed down generation after generation. One writer characterized it as the "most sadistic Storm of the Century." Since the blizzard began late Sunday morning, it had to compete with the magnet of televised National Football League playoffs. More than two feet of snow could not stop tens of millions from viewing the game. Football would not be denied. Masking the underlying worry of getting into work the following day created a powerful conflict for many New Jerseyans.

Ken Pileggi

Ken Pileggi was the branch manager and vice president of the Princeton, New Jersey branch location of SPT Electric Supply, a three-location, family-run electrical distributor headquartered in nearby Bridgewater.

There has been a long-standing statewide debate about whether Central New Jersey actually exists, as both North and South Jersey lay claim to the area. This debate rages to this day, a sensitive topic to many Central Jerseyans.

SPT had a long history of service to electrical contractors and residents in the area, as it was formerly Somerville Electric Supply. The previous owner, Irving Hirsch, had sold the business to Tom Knott and Jim Bulvanoski in 1984. Hirsch was a local civic and business leader who had been in business providing electrical supplies to the area for over thirty years.

Ken had listened to the television and radio forecasts in the days before the storm, with the blizzard warnings for all of New Jersey having been posted Saturday morning. SPT had never closed for snow, as its slogan, "The Customer Is King," promised twenty-four-hour service during emergencies. Its electrical contractor customers had around-the-clock need for supplies, particularly in bad weather. Tom Knott, the president of SPT, had a reputation of being the electrical contractor's best friend, and he had a no-nonsense approach to taking care of his customers. He was aware of the pending blizzard yet had not spoken to Ken since that Saturday morning.

Earlier forecasts had leaned toward heavy snow remaining well south of Bridgewater, yet forecast models and subsequent forecasts made an about-face. By Friday night, it became clear that all of New Jersey would receive one to two feet of snow, gale-force winds and drifts up to four feet. Governor Christine Todd Whitman declared a statewide state of emergency with a clear message to motorists: stay off the road.

> *The Cowboys were playing the Eagles in the divisional round of the NFL playoffs that Sunday, January 7. I drove over to my friend Pete's house to watch the game; the snow had started around noon, and it was already a whiteout, yet the roads were still manageable. Pete and I are lifelong Cowboy' fans, and we had a special interest in the winner, as we had tickets to the NFC Championship game in Dallas the following weekend if the Cowboys won, which they did, of course, 30–11.*
>
> *The beers were flowing, and the snow was pouring out of the sky. Three hours later, after the Cowboys' win, I had to drive home. As the weather got worse during the game, I made the decision to close the Princeton branch the following day. I called Jeff Wilson and our other employees from Pete's house and told them to stay home. About an hour later, Tom called me—he also knew Pete—and told me we were going to close the company on Monday. I told him I had already made the decision to close Princeton; he was not happy about that, and I really heard about it. Driving home—I lived in Hillsborough, about nine miles away—I couldn't tell how much snow had fallen, had to be about ten inches. Drifts had covered the backroads, and at times, I wasn't sure I was going the right way. I was becoming disoriented. Took me two hours to make a half-hour drive. As for Tom getting mad at me, I still think it's funny.*
>
> *The next morning, I opened the garage door to get the snow shovels to shovel our driveway. Me; my sons, Nicky and Tom; and our dog, Toffee, were going to attempt to do it. Toffee, who loved being outside, was so excited, she ran and disappeared into a big snowdrift. There was utter panic, tons of snow and no dog. For about thirty seconds, she was gone completely until Tommy went into the drift, found her and pulled her out. After that, she stayed inside. We ended up with thirty-one inches in the middle of our cul-de-sac.*

Dave Ryan

David Ryan was the general manager of Columbia Leather and Vinylide in Kenilworth, New Jersey. He lived in Freehold, New Jersey, in 1996 with his wife, Judy, a physical therapist, and their thirteen-year-old son, Danny. A lifelong St. Louis Cardinals baseball fan—even though his team was nearly a thousand miles away—he loved to use baseball references in almost any conversation.

Dave's father, Bartholomew, a native of Dublin, Ireland, was the amateur middleweight boxing champion of Ireland in 1949 and participated in the box-offs for the 1952 Canadian Olympic boxing team. He was one victory away from being an Olympian and possibly facing Floyd Patterson, a future World Heavyweight Champion, for a chance at an Olympic gold medal.

Up until that blizzard, I had always thought the Blizzard of March 1993 was the most brutal. We lived in Cranbury in '93, and on a Saturday morning, a foot of snow accumulated from early morning to around noon, then changed to heavy sleet and freezing rain. Compared to the '96 blizzard, which left my wife and son snowbound for two days, the '93 storm, they called it the "Storm of the Century," was like a Texas league Bud Harrelson single to left, compared to the Blizzard of '96, which was a true Mickey Mantle Griffith Stadium, 565-foot blast.

By 1996, we had moved to Freehold. I was watching CNN the day before the blizzard, a Saturday afternoon, and read the scrolling along the bottom of the newscast saying we could expect two to three feet *of snow. I thought it was a typo and they meant two to three inches. We had thirty-six inches of snow; I measured it with a yardstick in a portion of my yard that was very sheltered from the wind. Drifts were over four feet; our cars were just about totally buried. I took our dog outside without a leash to pee, and he vanished into a drift in seconds. It took several minutes to dig him out; he was a bit frozen yet OK. I couldn't get into work until the local roads were cleared and the state of emergency was lifted, and that wasn't until Tuesday afternoon. I had never before or since seen that much snow from a single storm.*

Meteorologist Joe Cioffi

Joe Cioffi is an Emmy Award–winning meteorologist who has been a familiar face on television and radio, covering the New York metropolitan area for thirty-five years. He has appeared on WPIX Channel 11, NBC Channel 4, News Channel 12, Long Island, and New Jersey 101.9 (radio) and currently has a regional and national audience with his podcast, *The Joe and Joe Weather Show*, cohosted with fellow meteorologist Joe Rao.

> *The winter of 1995–96 came early, with the first snow coming at the end of November. It seemed different at the start of that particular winter. I just had a sense this was going to be a different winter than most. We had gotten used to subpar winters when it came to snowfall in the metropolitan area. Mid-December brought the first one-foot-plus snowstorm for Long Island, and it left the area with snow cover right through Christmas into the new year. As the year ended, it seemed like something big was afoot.*
>
> *At that time, the long-range guidance was limited as to how far out it went, but as the new year came, it became obvious that a big weather system was going to impact the United States with a big storm. Model guidance began to zero in on this storm as it came in from the Pacific, and model run after run showed a potential for a major winter storm along the Eastern Seaboard.*
>
> *Everything began to come into place, with a big arctic air mass spreading out from Canada to the plains and Great Lakes. Low pressure formed in the Gulf states and started to move east-northeastward. Everything seemed to be in place now. All the weather models were consistent that this was going to be a one- to two-foot snow for most of the Northeast and north Mid-Atlantic states.*
>
> *The set-up was perfect by Sunday morning. The storm was in a favorable position. The high pressure to the north was holding in the Arctic air. It was a matter of sitting back and counting the inches. We were going to do nonstop coverage on News 12, Long Island, where I was working at the time. I got in the car at 8:00 a.m. for the thirty-minute drive to work from Centereach to Woodbury. The first flakes began to fall as I left the house. Within minutes, there was a coating on the ground. By the time I got to work, an inch had already accumulated. Temperatures were in the upper teens and lower twenties.*
>
> *I went into my office and put the graphics together. I decided to go eighteen to twenty-four inches of total accumulation, which was a forecast that was almost unheard of in that area, yet I felt certain of the outcome.*

Snow began to fall heavily at 10:00 a.m. at a rate of one to two inches per hour. By 4:00 p.m., as the first round of heavy snow had moved through, a foot was on the ground. The snow tapered off a bit yet was not done. The upper-air storm was pivoting around, and that put us in accumulating snow all night long into Monday.

I was on the air nonstop Sunday into the evening. I got to the hotel for a short sleep, and then it was back to the studio at 6:00 a.m. for the next round of coverage. It snowed all day Monday, January 8. The intense storm was off the coast of New Jersey, south of Long Island. It had intensified rapidly all night and began a slow crawl to the northeast. Finally, I got into my truck and drove home as the last of the snow ended at 5:00 p.m.

When I got home, none of the snow had been shoveled. The town plows had pushed a four-foot mountain onto my driveway, which made it very difficult to get into the house. I had at least two feet of accumulation. I got my snow blower and made very little progress in an hour's time. Exhausted, I was about to give up when a town truck came down the road. I knew the driver, and he used his truck to plow out my driveway. Thankfully, a saint was there to get the snow cleared, and I finally got into my house. I had a nice, hot meal and went to bed. It was one of the best experiences for me professionally with all of the television coverage I had to do.

Stan Kosinski

Stan Kosinski is a lifelong resident of the North End of Elizabeth, bordering Newark. He is a pressroom foreman for the *Star Ledger* and is also an administrator for the Sandy Hook Facebook group. Stan's father owned a bungalow on 3 Elder Street in Camp Osborn, a near-century-old sand beach bungalow community on Barnegat Beach Island on the New Jersey Shore. The Camp was buried under twenty-six inches of snow during the Blizzard of 1996, with drifts that nearly reached the roofs of the tiny homes.

I bought an old, full-size Chevy Blazer, I was working for Edwards Supermarkets in Elizabeth, which became a Stop and Shop. I was a shipping and receiving manager. Due to most loading docks in our area being the ramp down to building level, many trucks became stuck in the large drifts, delaying deliveries. I stayed many hours extra, not only waiting for deliveries but also using that big Blazer to ferry employees to and from work. We kept the milk, eggs and bread flowing.

Problems would persist after the storm, with no place to put the snow. Parking in Elizabeth was a nightmare. The city eventually used heavy equipment to clear the streets entirely so residents could park. They would truck the snow to the Elizabeth River and dump it. The Elizabeth hospitals were also looking for people with four-by-fours to get nurses in to work. There was no GPS or Map Quest yet; it was mostly handwritten directions. We had had our summer bungalow at Camp Osborn for only three years at the time, so I wasn't there for the storm, yet I heard drifts were halfway to the rooftops of many of the bungalows.

Michael Estrin

Michael Estrin is the owner of Estrin-Zirkman Sales, in Manville, New Jersey, in the heart of Somerset County in Central New Jersey. Manville received twenty-eight inches of snow, shutting down the historic factory town, which, for decades, was home to the Johns Manville Company, a global leader in the production of roofing and building materials. Most town residents were employed by Johns Manville, and the community was named Manville after the company that was the economic lifeblood of the town.

Estrin, a manufacturer's representative for Kichler Lighting, was to fly to Dallas to receive an award from Kichler, a landscape lighting company, as sales representative of the year, a nationally prestigious industry lighting award. As the blizzard threat grew with each forecast revision, Michael's travel plans were in peril. He had followed the radio and television forecasts as it became clear that his scheduled Sunday afternoon American Airlines flight to Dallas would be canceled, as the blizzard would be well underway, assuring widespread flight cancellations.

I was to fly to Dallas on a business trip. I got in front of the blizzard and called American Airlines to change my flight—was on hold for three and a half hours, took a nap and finally got the flight rescheduled. Got to Dallas in the nick of time to pick up my award as Kichler Lighting sales rep of the year. I had owned our lighting agency for only three years at the time, and the award was a major recognition for us. Then my wife called and told me she had taken our little boy, who was three years old, into the backyard to see the snow, and she turned her back and he instantly had gotten himself buried in a drift. He loved it, yet that ended playtime in the blizzard.

Craig Benner

Craig Benner, an HR assistant for Turtle Inc., an independent, century-old nationwide electrical distributor headquartered in Clark, New Jersey, remembered a chaotic Sunday afternoon as the blizzard arrived and intensified, making vehicular traffic virtually impossible.

At the time, Craig was employed in a retirement center, and the approach of the powerful nor'easter posed logistical problems for both employees and senior residents of the facility.

> *Oh, that was a fun storm. I was working at Kirkland Village in Bethlehem, Pennsylvania, a retirement center, that Sunday for my breakfast/lunch shift. By Monday morning, there were two feet of snow on the ground and giant drifts. Several of my coworkers were stranded and had to stay in some of the empty rooms. I was able to stay with a resident who had a two-bedroom apartment in the independent living center.*
>
> *As we did have limited staff, we switched to using paper plates and plastic silverware inside of the normal service ware, and the residents hated it, as it was an upscale living center. By the time I tried to leave, there was so much snow it was blowing through the heater vent in my car. I couldn't move an inch. I ended up being stuck there for three days.*

Meteorologist Christopher Stacheleski

As the Regional Cooperative Observing and Climate Services program manager, National Weather Service, eastern region, Bohemia, New York, Christopher Stacheleski oversees nearly 1,400 cooperative weather stations in the region and overlooks snow measurements in the station at all climate sites. He has been with the National Weather Service since January 2007 and has been involved directly with the climate and cooperative weather programs for years.

> *The significance of the Blizzard of '96 in Northern and Central New Jersey and metropolitan New York City cannot be understated over twenty-five years later. It was the biggest snowstorm overall for the area since February 1983. The '96 blizzard largely helped catapult the seasonal total to an all-time record at both Central Park and Newark. The snow total from that storm accounted for a significant part of the seasonal total.*

The other thing, historically, that is often forgotten with a lot of the subsequent blockbuster snowstorms is the depth of snow on the ground after the Blizzard of '96. It hasn't been equaled since in North Jersey. It still stands as the record at Newark Airport. But even though it's not an all-time record for the state of New Jersey—that's fifty-two inches back in 1961—there were large portions of northwestern New Jersey in the hill country that had near to over three feet of snow on the ground. There was still snow left from mid-December 1995 and an earlier storm in January 1996 on the morning of the blizzard.

Newton officially reported a forty-inch snow depth after the Blizzard of '96. Nothing since in official records equals that. The visual impact of the depth of snow on the ground from the blizzard and any old snow was staggering for a good three to four days, even after the storm ended. When you think about the fact that it's been nearly a generation since that much snow has been on the ground in New Jersey, I think it speaks volumes about not only that storm but that winter in general.

A lot of the more recent bigger storms had spots that had two feet of snow or a bit more on the ground following a single storm, but there was either no snow before the event or whatever fell melted or compacted quick enough it couldn't compound with subsequent events to a depth high enough to equal what was seen in January 1996.

It was the marquee meteorological event of one of the more epic winters ever, a storm that, if you experienced it, still stands out in your memory to this date. You might not remember just how much snow fell, and there was a lot of it. That it stranded you in place for a time. And then you had to dig out. In some places, the storm laid the foundation for another significant weather event in the subsequent weeks. It was the Blizzard of '96. Even though many areas having been impacted by it have seen bigger snowfalls in years since (February 2003 from the Presidents' Day blizzard and the Blizzard of February 2006, the Boxing Day Storm of December 26–27, 2010, the snowstorms of February 2015 or, more recently, in January 2016), this storm is widely viewed as the big one in the modern history of East Coast snowstorms of the last twenty-five years.

It digs up memories akin to those from the Cleveland Superbomb in January 1978, the Blizzard of '78 in New Jersey, metro New York City and New England. Snow flurries even fell as far south from this system as Florida, stretching from Tallahassee to just north of Tampa in New Port Richey, sending a touch of winter to a place many go to escape it. Yet the bigger snows the storm was noted for fell much farther north, reaching as much as forty-eight inches in Snowshoe, West Virginia. In the most severely impacted areas, transportation was crippled for days, municipalities struggled to clear the snow and find places to put it, mail service was briefly halted from Atlantic City, New Jersey, to Albany, New York.

Numerous roofs caved in from the weight of the snow, damaging structures. Schools and businesses closed for days, and newspaper delivery—along with other supply delivery—was halted. Many people were stuck at roadside rest stops. Along the coast, flooding occurred, and the onshore flow resulted in the bizarre site at Atlantic City, New Jersey, hundreds of clam shells washing ashore into mounds on the beach. The snowpack in many areas reached over two feet and, with additional snows and cold following the storm, laid on the ground until a warmup later in January with a heavy rainfall that took place and resulted in major river flooding.

ALAN SCHUTZ

Alan Schutz, the owner of Ascend Sales Group and Marketing LLC, a manufacturer's representative for ALG and Light Efficient Design, lived on Malone Avenue in North Edison, New Jersey, having lived there with his family since 1989.

An avid outdoorsman who challenges many of the wooded hiking trails in western New Jersey, Alan is a good neighbor and respected businessman who doesn't mind a snowstorm.

I had gone through several snowstorms in this home, including the 1993 March blizzard. Snow warnings were issued Saturday for the storm, expected to begin Sunday morning and last into Monday with up to two feet predicted. I usually left my car, a four-wheel 1996 Ford Explorer, Eddie Bauer edition, outside for snowstorms.

This time, I backed the car into the garage, knowing we were going to get a lot of snow. I wanted the best opportunity to get out of my home in my four-wheel drive. We didn't have a snowblower, since our driveway was fairly small; it was a two-car [wide] *driveway capable of 1½ cars deep.*

Backing into the garage proved to be extremely helpful. By Monday, we had over two feet of snow, and drifts were over six feet in some parts of our driveway. Using the Explorer as a snowplow, I was able to push large amounts of snow into the street on my side of the driveway in a small amount of time. I needed to get to work, yet my wife didn't, so I focused on my side. After I pushed the snow, I began clearing my driveway to the bottom using a shovel.

My neighbor, watching me, walked over to my house and began screaming at me because I had put so much snow into the street. He was frantic, and I assured him I would move the snow onto my lawn before I left my house for the day. I shoveled the snow onto my lawn, and he and I never spoke again. Until I moved out of North Edison in 1999, I avoided him like the plague. The memory of the insulting words he said to me caused me to lose all respect for my neighbor.

Snowstorms bring out the best—and sometimes the worst—in people. I am glad this happened in 1996. It was great that I cleared my side of the snow in my driveway to the ground; the untouched side of my driveway froze overnight, and it took a few days to clear it.

I did buy a snowblower shortly afterward and barely used it over the next few years.

Meteorologist Craig Allen

Craig Allen has been the voice of WCBS 880 AM for forty-two years. He is the longest-running chief meteorologist on a radio station in the country, according to CBS News. Allen worked on CBS until September 2006 and was the weekend meteorologist for WPIX Channel 11 from 2010 to 2020. He also worked part time for WNYWTV. He has also provided forecasts for Audacy.com and is currently a weekend First Alert forecaster for CBS News, New York.

I never got to witness much of the storm; it was constant on-air work, and I barely had any time to look out the window. It was an "all-hands-on-deck" type of storm, like I have been doing most of my life. I had to cover both radio (CBS 880) and television (CBS 2). Both radio and television coverage started an extra hour early that day. There was no way I was going to drive in, so I packed a suitcase and managed to catch the LIRR at the Freeport station around midnight. They were not following any schedules, just making every stop and all the local stops. So, it was a long ride.

Those were the days of the M3, a real workhorse train, so it plowed its way along at no more than thirty miles an hour, and the shoes had a hard time keeping contact with the third rail. Plenty of arcing and sparks had the ventilation and lights flickering or even out for long stretches of the ride.

Penn Station had a fair amount of people at that hour trying to navigate to their destinations.

Since the CBS broadcast center is on the West Side at Fifty-Seventh Street, I had a choice of the 1, 2, 3 or A, C, E trains to get to the nearest stop, which is Columbus Circle. I think I took the E, since I was in the front of the train and just continued walking to the West Side of Penn.

When I emerged to street level, I was in awe at the silence of the city. It was about 2:00 a.m., and I heard nothing but the wind as I walked, along with the sounds of shoveling from doormen and maintenance workers. I do vividly recall walking south toward Fifty-Seventh Street and west toward Tenth Street. Sidewalks already had a foot of snow on them, so I walked in the middle of the street as plows valiantly tried to keep the street accumulations to a minimum. My suitcase did not have wheels. It was strapped to a traveler's dolly cart, yet there was no way that it was rolling through the snow. I had all Fifty-Seventh Street to myself, no cars, buses or trucks, just one or two cabs may have rolled by, slightly faster than I was walking.

The hotel where we (the broadcast team) were staying is right on the corner before the broadcast center. I checked in, got dressed and it was off to work. There was nothing but on-air work. Radio on one side of the weather office, then across to the other side of the office to be on television. We had weather producers helping as well, so the forecast was as fresh as could be, with updated snow totals.

By then, the storm was in "nowcasting mode," and my job was with radar. I'm not sure who the other meteorologist was in the studio. I stayed in the weather office next to the big old monitor. I don't even remember if I stayed another day or caught one of the few available trains to get home that night. When I got home, it wasn't over. I had a lot of shoveling to do!

Frank Infantino

Frank Infantino and his wife were living in Staten Island in November 1995, awaiting the birth of the first of their three daughters, Cassandra. A Staten Island native, Frank spent his summers following his beloved Yankees. An electrical distribution salesman, he was looking for a new job and hoping to relocate to New Jersey in early 1996.

It's just a small piece that sticks out in my memory because of the situation my family was in at the time. I had just left Coleman Electric in November

1995 and was spending some time off sending out resumes, hoping to move to New Jersey. My wife was nine months pregnant with our first child. I figured what better time to take some time off, and hopefully, I could begin the new year with a fresh start at a new company. Cassandra was born on December 8, and the three of us lived in a new townhouse in Staten Island that we had bought the year before. We knew it was supposed to snow, yet the forecasts were all over the place and we didn't expect a blizzard. Well, obviously, with a newborn in the house, we had to make sure we had everything we needed for the baby. Neither of us were working at the time, so we didn't have anywhere else to go.

I remember being really nervous with the baby and what we would do if something went wrong. We had nearly thirty inches of snow, and it was nerve-wracking to be new parents with a newborn and be trapped in a house. I remember opening our front door, which, in this house, was the only door, no screen door in front, and seeing an exact replica of the door, a full seven feet high, imprinted perfectly on a snowdrift, doorknobs and moldings included, staring at me. The house was one of twenty-four on a small cul-de-sac, where every house had one small driveway and nowhere to put the snow. Every resident had to construct these monstrous, eight-foot-high snow walls to get in and out, like a giant maze, leaving barely enough room to navigate our cars into a narrow driveway.

ANTHONY "TONY" MAGLIO JR.

Maglio Electric is a second-generation electrical contractor that has been servicing Hunterdon, Somerset and Warren Counties in western and northwestern New Jersey since 1955. Tony Maglio Jr., like his father before him, provided electrical service around the clock for residents and businesses, including the annual Somerset County 4-H and Hunterdon County Fairs. Tony is a civic and community leader and has a unique story about the Blizzard of '96.

The forecast kept changing, yet by Sunday, we knew we were in for it. The snow began by noontime, and by sunset, the roads were mostly impassable. The drifts were four feet high and my crew could not make it in on Monday, so the business was shut down until the storm passed and people could get around again. It took us four days before we could get plowed out.

I live on top of Jug Town Mountain in Bethlehem Township. The county road leading up the mountain couldn't be cleared by regular snowplows; the snow was just too deep. They had to bring in D9 bulldozers to get the snow out of the way so people could drive. We were stuck inside for four days, and our business was shut down because the guy couldn't get up the hill to the house to pick up the trucks so we could get to work. That's all I can remember about that, but it was a hell of a storm.

There is a lot of history about Jug Town Mountain; that is where the moonshiners ran their stills back in Prohibition and the Depression. The last still was closed in 1955 in order to open up the road to the top of the mountain. The name Jug Town came from Jug Tavern, which dates back to before the American Revolution, to 1761. They used to mine ore up there, too. It's a part of the Musconetcong Mountain Range, and a lot of tourists visit there.

AL MUGNO

The king! Al Mugno has a lifetime passion for weather. "It all started when I was a young boy with Hurricane Belle in 1976 and then the winter of 1976–77, where we experienced many ice storms in my area. I have been a high school teacher at Northern Highlands for the past thirty-two years and started my teaching career at School 14 in Jersey City. I teach engineering and architecture at Northern Highlands and advise three clubs that I started—STEM, Future Architects and Engineers and the Weather Club."

During his life in Garret Mountain Reservation in West Paterson, New Jersey (now Woodland Park), Mugno witnessed winter's fury in the blizzards of 1978, 1983, 2003, 2010 and 2016. None of these tempests rival "the king," as Al, a longtime member of the North Jersey Weather Observers, remembers the monster Blizzard of '96.

I remember at work, a history teacher, Dennis Rouse, the day we came back from Christmas and New Year's break, saying to me on lunch duty that a good friend of his who knows a weather guru said we are going to have a major blizzard in about a weeks' time. I questioned, "Who?" And he said, "I can't tell, but he is a Hudson Valley weatherman."

I was intrigued by this. I watched Paul Kocin on the Weather Channel religiously. He was on and said it's going to be something to watch to see how this storm unfolds. That Friday night, we had a surprise birthday party for a good friend, and on the TV, they had the Weather Channel on and they showed what was going to be an intense storm, an all-out blizzard that could be life-threatening starting Sunday afternoon.

I turned to my friends and said, "We'll be out of school for a week if that happens." They said, "No way." That Sunday morning about 10:00 a.m., my mom called me at my fiancée's apartment and said it started snowing at home—much earlier than the early afternoon time frame that was reported on all major news outlets.

I was up in Nyack, New York, and said I'd be leaving shortly. As soon as I hung up the phone, it started snowing in Nyack. I got my things together and went outside, and to my surprise, the ground was covered and moderate snow and windy conditions had started. My thirty-five-minute drive would turn into a hellacious ninety-minute struggle. The Garden State Parkway had whiteout conditions in Paramus, and when I pulled up to the toll booth in Saddle Brook, the collector said Route 80 was closed and to be careful, the roads were becoming impassable. This was about two hours after it started snowing in this area.

I made it onto Route 80 and drove onto the center lane—or what I thought it was, for I could not see the road more than fifty feet ahead. A New Jersey state trooper with chains whizzed right by me. I was one of the only cars on the road. I prayed I would make it home OK, and I did. [There were] *whiteout conditions on the long Paterson stretch of overpass. Never again would I chance driving in such a storm.*

That night, the winds were howling; we had whiteout conditions, and the heavy blowing drifting snow was relentless. It was also frigid cold, I remember. The next day, it snowed almost all day long, and the plows once again got stuck on our hill. When they finally brought in the heavy-duty trucks to plow us out, they could not push nor move the snow off the road to the side.

There was so much snow that the plow had about ten feet of snow in front of its plow, pushing it to what seemed like nowhere. They made a clear path that was literally a car and half wide. My mom and her boyfriend walked to Route 80 on Tuesday morning. It was pristine, like it had not been touched! It was totally snow-covered, and drifts nearly covered the overpasses; you could not make out the road or overpass at Squirrelwood Road.

You would never know it was a major interstate, they said. When I went to work that Thursday, I saw something I was amazed at: they had what looked like Newark Airport snow removal equipment. They brought in snow throwers that started the job of clearing the roadway's right two lanes. There were two lanes open out of four in this stretch. Some exit and entrance ramps still were not cleared.

The cold that ensued after was rough, freezing solid anything that was not shoveled out. Drifts at our school covered entire entrances, and the parking lot spaces along the curb line were totally snowed under. A few days later, at night, I heard a rumble and looked out my window as a front bucket loader was picking up the snow from the street and dumping it on everyone's lawns to clear the roadways, which took five days to accomplish townwide.

12

THE PRESIDENTS' DAY BLIZZARD OF 2003

Arctic cold had invaded the eastern third of the nation in late November 2002, punctuated by an early winter coastal storm that delivered a half foot of snow to Central and North New Jersey on December 5. After Thanksgiving, November 28, fourteen of the next fifteen days saw below-normal temperatures in New Jersey.

The winter pattern favored repeated intrusions of cold air, which persisted through the Christmas holiday. On Christmas Day, New Jersey saw a white Christmas, as a sloppy and complex series of winter storms delivered heavy rain Christmas morning, which changed to moderate to heavy snow by late afternoon. Three to five inches covered most of Central New Jersey, and seven to eleven inches covered the northern and northwestern areas of the state.

January saw prolonged arctic air in place in New Jersey, with daytime high temperatures failing to reach thirty-two degrees for nearly two weeks, from January 13 to January 25. Two light snowfalls occurred during the month, between two and four inches, as the very cold winter persisted. Long-range computer models were signaling the potential for a strong coastal low impacting the Mid-Atlantic and Northeast region in early to mid-February.

Perhaps on Presidents' Day, the day it likes to snow.

By the second week of February, very cold air had once again settled along the Eastern Seaboard, plunging south into the Carolinas. Between March 12–14, 1993, and January 21–22, 2016, a span of thirty-three years, there were six powerful snowstorms in New Jersey, four blizzards and two on the precipice of blizzard classification. As Presidents' Day weekend unfolded, it was clear that a punishing and dangerous blizzard was bearing down on the Garden State.

Low pressure, which originated in the southern tier of the Rocky Mountain states and began taking the classic track for winter storm development, moved eastward into the Tennessee valley, with rain and freezing rain over the southern states and accumulating snow from Iowa into Michigan. There was a major ice storm in Central Kentucky, with roads turning into sheets of ice, creating perilous travel conditions. Forecasters in New Jersey and the Tri-State area were in unanimous agreement: the primary area of low pressure would transfer its energy to a secondary low-pressure system off of the Virginia–North Carolina coastline and "bomb out," moving north-northeastward along the 40–70 benchmark, toward eastern Long Island, and rapidly intensify into a historic winter storm, taking dead aim at New Jersey.

Snow overspread New Jersey, making a slow progression from south to north, on Sunday, February 16. The axis of heavy snow, with widespread rates of one to two inches per hour, developed over Southern New Jersey Sunday afternoon and advanced into Middlesex, Union and Essex Counties by nightfall, overtaking all of Northern New Jersey by midnight. Gale-force winds, sustained at forty miles per hour and gusting to fifty-five, drove the snow into impassable mounds, obscuring cars and leaving thousands homebound.

By Sunday night, lower level warming and a more easterly wind component resulted in the snow changing to sleet in Central and Southern New Jersey and even plain rain over Cape May County before it changed back to snow as the powerful nor'easter pulled away on Presidents' Day, leaving in its wake widespread and extraordinary snowfall totals. Newark received 22.1 inches, Garwood 23.5 inches, 24 inches in Little Egg Harbor and 27 inches in Green Pond in Morris County.

The crippling blizzard resulted in the closure of major metropolitan area airports from Ronald Reagan Washington National, Baltimore Washington International and Philadelphia International to LaGuardia Airport in New York City. New Jersey Transit suspended bus service, although rail service, with delays, was maintained. With the blizzard occurring on the Presidents' Day weekend, schools and many businesses were closed.

This was a destructive and unsparing blizzard, in the top tier of ferocious winter storms to ever impact New Jersey. Governor Jim McGreevey declared a statewide state of emergency, employing 2,500 state workers who used

2,000 snowplows to clear New Jersey streets and highways. The weight of the heavy snow collapsed a roof at a Jobs Corps Academy Learning Center in Edison Township, resulting in one man, who was taking a cigarette break, losing his life, and four people sustaining injuries.

Dave Ryan

Dave Ryan, forty-six at the time; his wife, Judy; and their son, Danny, had moved from Cranbury, New Jersey, into their new home in Farmingdale on February 1, a Saturday. Dave was the general manager for Vinylide Coated Fabrics in Kenilworth, and Judy was a practicing physical therapist at International Physical Therapy, her business in Howell Township. February 1, 2003, was the date of the tragic explosion of the *Columbia* space shuttle, which took the lives of seven astronauts.

> *We had just moved into our newly built home in Farmingdale, the first of what would be six homes on the development. It was Saturday, February 1, as we had closed on the house January 31. I went into town for a haircut when I heard Big Joe Henry on NJ 101.5 interrupted his program and said, "It is reported that contact has been lost with the space shuttle* Columbia.*" It was one of those moments where you remember exactly where you were and what you were doing, like when the shuttle exploded, when President Kennedy was assassinated and September 11, 2001.*
>
> *Since the development was new, we were the only humans on the block, the three of us were isolated. It was a beautiful, wooded area, yet it was very unnerving. The silence was deafening. You couldn't hear anything. A few weeks later, on Presidents' Day weekend, we were waiting for the arrival of the big blizzard. That's all you heard from the media, nonstop coverage, up to two feet predicted. The snow started early Sunday afternoon, light for an hour or two, then very heavy snow and wind. It snowed nonstop for twenty-four hours. Danny, myself and Judy were stranded. We didn't even have a driveway yet. We couldn't leave the house for three days until the city plowed the temporary streets in the community. We got twenty-five inches of snow, and we spent a lot of time shoveling and watching TV.*

AL MUGNO

Al Mugno has been a high school teacher for thirty-two years, teaching engineering and architecture at Northern Highland Regional High School. He began his teaching career at School 14 in Jersey City and grew up on Garret Mountain in West Paterson, New Jersey. He has coached varsity soccer, track and field and basketball and currently coaches middle school boys' soccer in Waldwick. He was selected TEENJ teacher of the year in 2012.

He has been on Steve Adubato's *One-on-One* and featured on *Classroom Close Up* three times, News 12 and Fios 1 News for developing an award-winning project called the "Cardboard Regatta Boat Race" for his honors engineering course. The race was held at Crestwood Lake in Allendale each June, where his honors-level juniors and seniors design, calculate the buoyancy, build and race boats that are meant to hold them and their partner, paddling 150 meters through a course on the lake. He is a student of weather and winter storms and is a member of the North Jersey Weather Observers.

> *Traveling to work on Wednesday morning, days before the blizzard, the radio station had meteorologist Bill Evans from Channel 7 News on, and he said the storm could be biblical! I let out a euphoric "Yes" in excitement. The radio hosts said, "Wait a minute, Bill, did you just say biblical?" He replied, "That is exactly what I just said, and I believe this has all the ingredients to be one of if not the biggest storm in the New York, New Jersey metro region history!"*
>
> *I was dumbfounded. I thought the Blizzard of '96 was the* king. *So, I started telling my friends at work about this, and they were skeptical. I remember Paul Kocin that night on the Weather Channel saying this would be an impactful and dangerous storm from Virginia through Boston. The Saturday before the storm was to start, I went to the grocery store early in the morning, and it was jam packed. Milk and eggs were in refrigerators that usually were for other frozen foods due to the overwhelming demand.*
>
> *Bread was in rolling racks in a makeshift area in the front of the store, as were pallets of rock salt and boxes of shovels. That Sunday morning, I had the Weather Channel on and was hearing reports from family and friends in Southern New Jersey that it started snowing already. I watched the radar as I saw the precipitation marching slowly north.*

My wife's cousins in Middletown called in the late afternoon and said it started snowing around 1:00 p.m., and they had a couple of inches already. You could see all the virga on the radar, and it was somewhat upsetting because we were losing the overrunning precipitation to all the cold, dry arctic air. At 6:00 p.m. my in-laws came by, and they were on their way out for dinner. I said the snow should start around 9:00 p.m., and once it does, it'll come down heavy. My father-in-law spoke to his sister in Edison, and they also had several inches.

Where was our snow? I was getting frustrated. Then at 9:00 p.m., it started light, and at 9:30-ish the skies opened up, and it started to really come down. In minutes, there was a heavy coating, and the winds started to kick up. My in-laws left for dinner in a hurry; their usually ten-minute trip to the restaurant took twenty minutes.

I stayed up with our backyard spotlights on and watched the peaceful snowfall whip around our backyard. We went to bed around 11:00 p.m., and there were several inches on the ground, about two and a half inches at that point already. I watched the radar on the Weather Channel and the heavy precipitation was about to swing through overnight.

I woke up the next morning around 8:00 a.m. and could not believe my eyes! The cars had drifts that covered them; our minivan had a drift that covered the whole front of it. And it was a full-on blizzard still. Our young kids got up, and the snow on our back sliding door from drifting was over my youngest son's head! We still had about eight more hours of snow to go! At about 2:00 p.m., I decided I was going out to start the clean-up. I had to climb out our back room window because the snow drifted up against both our back door and front door that would not open! The snow was halfway-plus up both doors; I could not open it. I trudged my way to our unattached garage and opened the door. I started our snowblower, and it was struggling. The twenty-four-inch height was not enough, as snow was over the top of it as I tried to snow blow.

It took me a couple of hours to do our small single-lane, two-car driveways; then I went and helped my elderly neighbors as well. An incredible storm—I measured twenty-six inches around my property and drifted six to eight feet high throughout our neighborhood.

13

THE SURPRISE BLIZZARD OF '06

The arrival of the new millennium marked the end of thirty years of winters with largely below-normal snowfall in New Jersey. From 1970 to 1999, there were only four snowstorms with accumulations of twelve inches or more, three of which were classified as blizzards by the National Weather Service. For three decades, there were only six winters with above-normal snowfall at Newark.

On December 30, 2000, a comma-shaped, tightly wound and intense coastal storm buried most of New Jersey and New York City in twelve to eighteen inches of snow, with screaming winds and drifting powder. Unofficial snowfall totals of twenty to twenty-six inches were reported over northern and northwestern counties. Resembling a winter tropical storm, the storm did its damage in ten hours, leaving New Year's revelers with a wintry memory.

The December 2000 snowstorm not only heralded the arrival of a new millennium, but it also marked the beginning of two decades of winters with near- or above-normal snowfall most years. There were four consecutive winters in the 2010s in which 40 inches or more of snow were recorded at Newark Liberty Airport. Between December 2000 and February 2021, seven prolific snowstorms occurred over New Jersey, including the great Blizzard of January 22–23, 2016, which established a record 27-inch accumulation at Newark Liberty Airport and 27.6 inches at Central Park in New York City.

The winter of 2005–06 began with two early season snowfalls, totaling a combined nine inches at Linden and Newark, providing a wintry start to the holiday season. The last two-thirds of December and January were unusually mild, with less than an inch of additional snowfall as the calendar turned to February. A mild, zonal jet stream directed Pacific air across the nation and kept temperatures well above normal over the Eastern Seaboard. The first week of February was even milder, with a springlike sixty-one degrees in Elizabeth on February 3 and sixty-four degrees at Somerville.

Then, with little advance warning, winter roared back. A confounding and crippling winter storm left its mark in the record books and in the memories of millions of New Jerseyans.

Colder air had settled into the Northeast during the second week in February, a midwinter pattern change that replaced the mild temperatures with winter cold. A weak area of low pressure developed over Alabama and Georgia early in the week of February 5, bringing rain and snow from West Virginia into Virginia and Maryland by late Friday and into early Saturday, February 11. Although the disturbance was expected to undergo rapid intensification off the southern Mid-Atlantic coastline, the snowstorm potential for New Jersey was not underscored by most forecasters, as model guidance consistently kept accumulating snow from the developing nor'easter to the south of New Jersey.

As late as Wednesday evening, longtime Accu-Weather meteorologist Dave Bowers, in his revised evening forecast for Thursday into Saturday, said, "By Saturday night a storm to our south may bring a small accumulation of snow to New Jersey and New York City." Then everything changed!

Forecast models abruptly changed course from the guidance provided earlier in the week. Where the threat of significant snowfall in New Jersey was not embraced by forecasters through Wednesday, now, the likelihood of a major snowstorm, possibly reaching blizzard proportions, was making weather headlines on radio and television. By Friday evening, the winter storm watch that had been issued earlier that morning for Central and Northern New Jersey had been upgraded to a blizzard warning. Now, sixteen to twenty-four inches of snow were forecast for the New Jersey suburbs close to New York City and the five boroughs.

The now-massive shield of snow, with some rain mixed in at the start, overspread New Jersey Saturday afternoon. By late Saturday night, it was clear that a historic snowstorm was unfolding. Snowfall rates of one to three inches per hour occurred in heavy snow banding, with total whiteout conditions punctuated by loud cracks of thunder after midnight. The heavy snow pounded the region until early Sunday afternoon. Winds gusting to over forty miles an hour buried cars and trapped motorists on highways.

Newark measured 21.3 inches of snowfall, Rahway 27 inches, Roselle 24.6 inches and Hoboken 20.7 inches. An all-time snowfall accumulation record was shattered in Central Park in Manhattan, 26.9 inches, much of it falling between 5:00 and 11:00 a.m. Coastal Southern New Jersey was largely spared, as the precipitation fell as rain through most of Saturday night before changing to snow early Sunday, with snowfall totals of five inches or less.

There were $5 million in damages recorded from Virginia to Atlantic Canada, yet there were only three confirmed fatalities. Fifteen thousand New Jersey residents lost power. The snowstorm occurred on the weekend, yet over half a million households, from Washington, D.C., to southern New England suffered power outages in the path of the storm—a winter storm for the ages.

Frank Infantino

Frank Infantino, a native of Staten Island, had moved his family, which included three young children, to Central New Jersey in the mid-1990s. An inside salesman for Turtle & Hughes, a New Jersey–based electrical and industrial distributor with a branch location in Bridgewater, Infantino lived in Bound Brook, Glen Gardner and, as the Blizzard of 2006 arrived on the weekend of February 11–12, 2006, nearby Somerville.

> *We survived big snowstorms before, the Blizzard of '93 and Blizzard of '96. In '96, we were trapped in our home for days. I was a new father, and the snow drifted up over the door. It was the second weekend in February 2006, and there wasn't much warning about a blizzard. Usually, it's all you hear about for days. We were living in Somerville at the time, about ten minutes from my job. At work on Friday, Tom, an inside salesman who was a weather nut, sent out an email to management and the salesmen about a winter storm warning for eighteen inches of snow Saturday night and Sunday.*
>
> *It began raining and snowing a bit by Saturday afternoon and changed to snow by sunset, and it was really coming down. I was staying in Somerville with my two daughters and my son, who were all little at the time. Stuck inside during a blizzard with three young children is a test of parenting skills. Saturday night was nonstop heavy snow, thunder and lightning. It kept getting worse and worse. Before midnight, we had over half a foot with*

lots of drifting. The car was totally covered, it was '96 all over again. Next problem was there wasn't much food in the house.

The following morning, Sunday, it was still pouring snow, at least fifteen inches by mid-morning. The only store open was the 7-11 on Mercer Street, so I had to trudge a half mile down to the 7-11 to buy us breakfast. On top of that, we had to work the next day. Our manager didn't believe in closing for snow, not even a blizzard.

14

THE CHRISTMAS WEEKEND BLIZZARD OF 2010

In Monmouth County, New Jersey, snowdrifts of up to five feet contributed to stalling a passenger bus on the Garden State Parkway, where snowplows were having a difficult time cleaning because there were so many stranded cars cluttering the ramps," state police spokesman Steve Jones said. Ambulances couldn't reach the bus, and state troopers took their own water and food to the bus to give to people who were feeling ill.

There have been eight powerful nor'easters classified as blizzards or near-blizzards by the National Weather Service that have unleashed their fury on New Jersey since 21.4 inches of snow was measured at Newark Airport on December 11–12, 1960. There have been dozens of major snowstorms, yet what defined these eight winter storms is that they all dropped twenty inches of snow or more on parts of the Garden State.

Every winter storm is different in character, and that is one of the reasons they are very difficult to forecast. If very cold air is in place, will it serve to keep the snowstorm far south of New Jersey, which happens a few times every winter? Will the storm track close to the coast or over interior New Jersey, causing any snow to change to rain, meaning highway flooding instead of snowplows?

The winter of 2010–11 started uneventfully. There was a cold snap around the middle of December, and only a tenth of an inch of snow had fallen at Newark that month. Then as Christmas week arrived, a winter storm scenario began to slowly unfold. Christmas fell on a Saturday, and as early as Monday and Tuesday, forecast guidance models were in agreement that a developing area of low pressure from the Gulf of Mexico, originating from two weak low-pressure areas that dropped southward from Canada, would track eastward over Florida and make a sharp left turn up the Eastern Seaboard.

The snow would arrive in New Jersey early Sunday, Boxing Day in Canada, a day too late for a white Christmas. As the holiday weekend neared, the European, Canadian and U.S. models were not in agreement. The GFS model saw the storm moving just off the coastline, supporting very heavy snow in New Jersey, New York City and into southern New England. The European model saw the powerful low-pressure area staying far enough offshore, sparing New Jersey a crippling snowstorm.

Everything changed Friday. Daytime model runs began shifting the axis of heavy snow northward, concluding that a storm of blizzard proportions was now a certainty for New Jersey and the Tri-State region beginning after midnight on Christmas and continuing into early Monday. Winter storm watches were issued at 4:00 a.m. on Christmas morning and were upgraded

Drifts nearly six feet deep almost obscure a bungalow on Cummins Street in the Camp Osborn beachfront community during the blizzard of December 26–27, 2010. *Courtesy of NOAA.gov.*

to winter storm and blizzard warnings at 4:00 p.m. Christmas afternoon. Millions of people who were enjoying Christmas weekend had to prepare—and quickly!

It was a prolific blizzard, one of the top five in the recorded history of New Jersey. A record 34 inches of snow was measured in Brick, New Jersey; 32.5 inches at Edison; 21.3 inches in Linden; 32 inches in Rahway; and 20.0 inches in Central Park in New York City. Widespread 20- to 30-inch accumulations occurred throughout New Jersey, even close to the immediate shoreline, where amounts were somewhat lower due to sleet mixing in during the height of the storm.

Winds were sustained at thirty to forty miles per hour, gusting as high as sixty miles per hour, with mammoth drifts of four to six feet totally burying cars and leaving many residents struggling to even leave their homes. Many motorists were stranded on New Jersey state and local highways, although the Christmas weekend reduced the volume of traffic considerably. Signal problems resulted in a suspension of operations for New Jersey Transit and Amtrak trains between New York's Penn Station and New Jersey, impacting thousands.

Another blizzard footnote was the seeming indifference of New York City Mayor Michael Bloomberg, who was said to have been vacationing in Bermuda during the blizzard. On Monday, Bloomberg had urged New Yorkers not to panic. He noted that people should be patient, stating that "Broadway shows were full last night."

The blizzard ground New York City to a halt. Roadways would not be cleared until rush hour on Tuesday morning, prompting Bloomberg to tell commuters to "be a little patient, it's a bad situation," as the sluggish pace of snow removal had angry New Yorkers remembering the Lindsay snowstorm forty-one years earlier.

New Jersey Governor Chris Christie was roundly criticized for vacationing in Disney World while Lieutenant Governor Kim Guadagno spent the holiday in Mexico as millions of New Jersey residents were buried in the historic snowfall. Acting Governor and State Senate president Steve Sweeney issued the state of emergency for New Jersey Sunday night, as 125 accidents and 1 death had been reported during the height of the powerful winter storm. During the winter of 2010–11, 56.7 inches of snow fell at Linden, making it one of the snowiest winters in New Jersey history.

Meteorologist Joe Martucci

Meteorologist Joe Martucci thrives on the leading edge of media in weather. A certified broadcast meteorologist and certified digital meteorologist with the American Meteorological Society, he is the only person in the state with both certifications. Currently the chief meteorologist for Triax 57, Martucci was born and raised in Union Township, New Jersey, graduating with a meteorology degree from Rutgers University.

A ten-time New Jersey Press Association (NJPA) Award recipient, Martucci has covered New Jersey weather, particularly Jersey Shore forecasting, for over a decade. He is the owner of Cup A Joe Weather and Drone and previously worked a six-year stint at The Press of Atlantic City and Lee Enterprises. Joe also freelances for the News 12 Networks. He has been featured in *The New Yorker*, *Deadliest Catch* and Rutgers University publications.

The Boxing Day Blizzard of 2010 was the biggest snowstorm in my life to this day. It was also the perfect combination. I was an aspiring meteorologist at Rutgers University. It was winter break of my sophomore year, back home in Union Township in Union County, and I was able to understand the meteorology of nor'easters more.

The northern and southern branches of the jet stream phased together perfectly. There was a very strong jet stream up the East Coast, which was allowing the surface low-pressure system to strengthen quickly. That Christmas Eve and Christmas Day, before the blizzard arrived, was cold. No rain-snow line here. It was just a matter of how much precipitation the low-pressure system would produce, which would fall as sweet snow. Turns out, eastern Union County would be the New Jersey bullseye for this nor'easter.

With the Christmas holiday and the snow coming the next day, it was the best gift you could ask for. Christmas Day was at our house. Along with the "What did you get or give for Christmas?" was plenty of chatter about the impending storm. I think I even put out a forecast post on what was still a young Facebook for my friends. I think the forecast underperformed expectations.

Snow began that Sunday morning. It snowed for about twenty-four hours straight. In our rec room, we had a sliding glass door out to the back deck. First, the snow was higher than our dog, Bella. I'd go outside every so often to measure the snow in the backyard. You could barely see three houses down at times, and we all had fifty-by-one-hundred-foot lots.

A tree amid a backdrop of suburban homes reflects the snowfall depth of the Christmas weekend blizzard of December 26–27, 2010. *Courtesy of NOAA.gov.*

Eventually, the snowdrifts were halfway up the sliding glass door. The snow was so high, you'd have to march through it, lifting your legs up and down. My snow boots would get caught in the thick snow coming up. You couldn't even tell where the three steps were going down the back deck!

Fifteen, twenty, twenty-five inches of snow. You couldn't believe it. I was submitting my storm reports to the National Weather Service, hoping to make it on the local news. I don't think I ever did, as the storm total in Elizabeth was always an inch or two higher.

My final total was an exact 27 inches. However, clearing the snowboard and obtaining an accurate measurement in this wind and snow was always a challenge. Elizabeth ended up with 31.8 inches of snow. There was dispute as to whether that total was completely accurate though. Either way, it was an incredible snow for all of us.

My mom always says she wants a big snow on Christmas Day. Santa was only a day late on her wish.

Meteorologist Nick Gregory

Fox 5 Chief Meteorologist Nick Gregory is the longest-serving television meteorologist in the Tri-State area. He joined FOX 5 News in December 1986. Previously, Gregory served as the morning meteorologist for CNN in Atlanta. Prior to CNN, he served as a meteorologist for WTLV-Channel 12 (ABC) in Jacksonville, Florida.

He has been on the American Meteorologist Society's Board of Broadcast Meteorology and was chairman of the board for this group in 1992. Gregory is the recipient of much recognition during his long career. He received the AMS Seal of Approval for Excellence in Television Weathercasting. The *New York Post* and the *Daily News* have named him the most accurate weather forecaster in New York City.

A 1982 graduate of Lyndon State College in Lyndonville, Vermont, Gregory holds a bachelor's degree in meteorology. Gregory is an FAA-designated pilot examiner, authorized by the FAA to conduct flight tests on behalf of individuals who want to become pilots.

Gregory is also a licensed pilot and flight instructor and is featured on any FOX network aviation feature story. Very active in philanthropy, Gregory has flown missions for Angel Flight, which provides free transportation for financially needy families and patients who need to travel to receive treatments.

Gregory and his wife, Athena, are the parents of three daughters, Anastasia, Valia and Ariana.

The one common element you'll find with meteorologists is that they have a love and passion for snowstorms. I'm no exception to that, as I have been enamored by snowstorms since I was a young boy. Since I've spent most of my life in the metropolitan area, I've had the chance to experience, study and forecast nor'easters and blizzards. This goes all the way back to the 1969 Mayor Lindsay snowstorm up to the January 2016 blizzard, which gave Central Park its record snow accumulation, 27.5 inches.

Covering these storms can be quite exciting but also quite tiring. For example, the January 7–8, 1996 nor'easter brought twenty inches of snow or more to parts of the city suburbs and Northern New Jersey. As the chief meteorologist at Fox 5, I was called in to cover the storm over a period of three days. The TV station put me up at a nearby hotel so I didn't have to deal with traveling back and forth from home. I was also broadcasting morning weather reports at the time on radio station 106.7 Lite FM.

The blizzard of December 26–27, 2010, began burying vehicles as heavy snow overspread the Mid-Atlantic and Northeast. *Courtesy of NOAA.gov.*

Between the radio broadcasts and the television coverage of the blizzard, I slept a total of ten hours over three days.

Since I'm usually called into Fox 5 to cover a snowstorm, I was fortunate that the February 11–12, 2006 blizzard struck on a weekend. The station wasn't planning extensive coverage, so I had a rare opportunity to witness the storm from home. We received twenty-eight inches of snow, and I was ecstatic to watch it happen. That ecstasy would later fade away when it came time to clear all that snow. For years, I relied on my trusty shovel (and some help from my family) to dig us out of these monster storms.

I felt I could do the same this time. I was wrong. While I did get some help, there was so much snow that I was left holding the shovel to finish the job. Nine hours later and two changes of clothes, I was done. As I proudly took once last look at my now-cleared driveway, I sauntered into the house, guzzled down two glasses of water and proceeded straight to my laptop. A few clicks on Sears.com would then bring a twenty-nine-inch-wide snowblower to our home a few days later. After its first use, I was dumbfounded why I hadn't purchased one sooner. It's been a relative pleasure to remove the snow ever since.

Another storm that stands out in my mind is the December 26–27, 2010 blizzard. Our morning guy was away on vacation, so they decided to have me come in that Monday morning for storm coverage. This time, though, I wasn't going to be put up at a hotel. Knowing that I would have to arrive early in the morning, I went outside before going to bed to clear a portion of the driveway. Our middle daughter came outside with me to help in what was the height of the storm. The wind was howling at around forty miles per hour, and the snow was punishing against our bodies. She shouted out, "Have you ever experienced anything this bad?" I replied with an emphatic "No!"

I hopped into my Chevy Suburban around 4:00 a.m. so as to be at work for the 6:00 a.m. broadcast. Expecting the driving conditions to be awful, I was pleasantly surprised that the interstates were in decent shape. That would all change upon entering Manhattan. This became known as the Mayor Bloomberg storm, since he was away in Bermuda and many streets of Manhattan were unplowed, making for a hard commute for many.

While the avenues were drivable, the side streets were awful, still covered with fifteen to twenty inches of snow. My trusty Suburban plowed through the fluffy snow like a warm knife slicing through butter. The only issue would be the large snow berms the plows had created the intersections. Once again came the Suburban to the rescue, as I would punch the accelerator and blast through the berm.

That worked well until I had to turn right onto Sixty-Seventh Street from Second Avenue. Emboldened by successfully traversing these snow berms, I decided to turn and gun it at the same time. Big mistake. This resulted in not enough momentum to break through the berm, causing the car to sink and get completely stuck. The snow pile was quite thick and extended up to the windows, making it impossible to open any door. My only escape was to exit via the tailgate.

My other mistake was not to have a shovel in the car. However, there were four men clearing the sidewalk of a nearby apartment building. I sheepishly approached them and asked if I could borrow a shovel. The main worker agreed but then recognized me as he politely scolded me saying, "Hey! You should know better than to do that!" We all had a good laugh.

All four men offered their help, and within about five minutes, we had dug up enough snow to see the tires. As I settle into the driver's seat, I make my plan. I'm going to put it into reverse to back off the berm then gun the car forward. Success! I blasted over the pile of snow, subsequently turning left and heading down into the parking garage. Upon exiting the garage, I walked over and gratefully gave each of the men twenty dollars and then proceeded to the TV station.

Despite all that drama, I made it on the air on time.

Al Mugno

Al Mugno, a teacher and member of the North Jersey Weather Observers, an organization of dedicated weather enthusiasts and meteorologists, knew a crippling blizzard was on the table for New Jersey. He also wanted a white Christmas for his two young sons. However, by Friday morning, the forecast verdict was not yet in.

Christmas Eve, I turned on the Weather Channel, and there was a winter storm watch posted in the morning. As we went to church and came home for dinner, the watch was canceled. I was unhappy because, for my young kids, there is nothing like the holidays with snow on the ground—just makes them that more special.

On Christmas Day, we were up at my brothers-in-law for Christmas dinner, and I went on his computer and looked up the AccuWeather site. It was about 8:30 p.m., and we just finished dessert. I said, "Holy smokes, they put up a blizzard warning for tomorrow afternoon and night." Everyone in our family said it was going out to sea and "you are pulling our leg." I said, "Look for yourself!"

My brother-in-law put on the Weather Channel; Paul Kocin was on saying, "This is not a joke." The storm is taking a dramatic turn, and the Megalopolis region from D.C. up through Maine can experience a pretty powerful blizzard. My wife's cousin was driving back to D.C. the next day, and he asked what time he should leave. I said, "Get on the road by 8:00 or 9:00 a.m., because it will start down south much earlier than up here." He left by 8:00 a.m. and got home in D.C. by about noon, and about two hours later, the snow started!

That day, we went over to our friends' across town to have some leftover holiday food and cheer and lounge by the fireplace. At 5:00 p.m. the snow started, and we left our friends by about 6:00 p.m. The roads, within an hour, were not good. We luckily got behind a plow and salt truck, and they helped us on our travels back home. We made some hot chocolate and watched the blizzard unfold outside our big bay window.

At around 9:00 p.m., the storm was cranking, and we had whiteout conditions. We could barely see our neighbors' houses about one hundred feet away! We went to bed and heard the wind howling. We woke up the next

The life-threatening, historic blizzard of 2010 left two to three feet of snow and whiteout conditions that imperiled travel throughout New Jersey. *Tim Armstrong, meteorologist, NOAA.gov.*

> *morning and looked out the window to see snowdrifts covering our eight-foot shed, and the snow was halfway up our back door!*
>
> *My oldest son, Mike, had to dig a path for our six-month-old puppy to go to the bathroom outside! My neighbor, who worked for the local town DPW, said later that they called all plows off the road because of the dangerous whiteout conditions from 12:00 a.m. until 4:00 to 5:00 a.m. I measured twenty-four inches for this storm, with drifts of eight feet–plus in our neighborhood. Up at school, entire entrances were drifted over, and one area had twelve-foot drifts covering the walkway.*

Meteorologist Kyle David

Kyle David is a graduate of the Rutgers meteorology program with a bachelor's degree in meteorology. At Rutgers, he participated in the American Meteorological Society student chapter, becoming active in field weather station maintenance and on-air forecasting. Kyle worked on campus at Rutgers and also served as a student intern for ABC 7 and a weather producer for Fox Weather. Kyle is currently pursuing a master's degree in communication through Johns Hopkins University to further refine his communications skills with diverse audiences.

He remains active in meteorology through weather consulting for small forecasting companies and works independently through multiple platforms and projects. Kyle's passion for meteorology, particularly tropical systems and severe weather, came from his mom and her longtime interest in weather. He tells the story of the December 26–27, 2010 blizzard through the eyes of a ten-year-old and future meteorologist.

> *I was ten years old when the Boxing Day Blizzard of 2010 hit New Jersey. It was one of many weather events in New Jersey that kindled my already-piqued interest in the weather. December 25, Christmas Day, I remember the buzz about the blizzard at the Christmas dinner table.*
>
> *December 26, Sunday, as the blizzard was beginning, I was at home with my family, still celebrating Christmas, and I remember that my father, who was working at Petco then, had to go to his store in Parlin because no one else working could get there to care for the animals before the blizzard blanketed the roads with one to two feet of snow. It was snowing when he left in the morning, and at the time, I was sad that he had to go to work. Now, I look back on that as a meteorologist and think,*

"That is the last thing you should do when expecting a major blizzard like the Boxing Day storm."

Nevertheless, it was commendable of him to brave the elements to go and take care of the animals at the store. While he was out, I spent time with my brothers, playing with our Christmas gifts and enjoying ourselves. My mom was preparing a nice post-Christmas dinner for all of us as the snow started to fall. My dad returned during the early evening hours, and we all celebrated Christmas with a nice meal as the snow began to really come down heavily. Snow was already covering the ground and really piling up; it was now difficult to see through the onslaught of big snowflakes. I woke up the following day to a fantastic sight for any ten-year-old: mounds of snow blanketing the neighborhood.

At first, I thought, "Wow, this is a lot of snow! I can build a big snow fort in this!" After a little while, I wondered what could've caused this much snow. It fueled my curiosity about the weather and to learn more about what happened and why it happened. I knew my future was as a weatherman.

15

THE GREAT BLIZZARD OF 2016

Beach erosion from Cape May northward to Long Beach Island displaced tons of sand, with nineteen of Jersey's sixty-six beaches sustaining major damage, to the extent that many would not open on time for Memorial Day and the upcoming summer season. The most drastic losses of sand occurred in the Holgate section of Long Beach Township on Long Beach Island, where beaches dropped twelve to fifteen feet in elevation, according to the survey.

During the weekend of January 24–25, 2015, the airwaves and social media were overwhelmed! Supermarkets were packed, lines extending out the door. Liquor stores were flooded with customers, and businesses throughout the New Jersey and New York metropolitan area, from southern Connecticut to Philadelphia, announced they would be closed on Tuesday, January 27. Massive preparations were made for what a large consensus of forecasters were calling a monster blizzard, with snowfall accumulations of twenty-four to thirty-six inches expected to bury tens of millions.

Governor Andrew Cuomo ordered the New York City subway system shut down for the first time in its over century-long existence. Mayor Bill DeBlasio declared that "this could be the biggest snowstorm in the history of New York City." In New Jersey, Governor Chris Christie declared a state of emergency as light to moderate snow began blanketing the Garden State by late Monday afternoon, resulting in a treacherous evening commute.

By Monday evening, the snow had tapered off to a wintry mist. The hours passed; the forecasts remained unchanged. Then, shortly before 11:00 p.m.,

The monument at High Point State Park in the aftermath of the blizzard of January 22–23, 2016. *Nick Stefano, New Jersey Weather Observers.*

the Weather Channel's continuous storm coverage told viewers that only twelve inches was expected in New York City. Most news outlets maintained their forecasts of heavy snow accumulations. People went to bed expecting a Tuesday morning nightmare. It never arrived!

The powerful low-pressure center moved fifty miles farther east than computer guidance was depicting. The heavy snow barely made it westward into the five boroughs. Central Park measured 9.8 inches. Farther west, into eastern New Jersey, the snow accumulations were smaller. Linden had 4.3 inches of snow. Newark had 6.5 inches. Only a few inches fell over interior Central and North New Jersey. Eastern Long Island and southern New England bore the full brunt of the blizzard, with a widespread twenty to thirty inches smothering the coastal communities.

The meteorologist in charge at the National Weather Service in Philadelphia apologized for the failed snowfall predictions. For forecasters throughout the Tri-State area, it was a stern reminder that Mother Nature has the final say. The 2015 "blizzard that wasn't" served as a learning tool for forecasters. Winter storms have a degree of uncertainty. There are many moving parts and often conflicting information, and a change in any one of these parts, such as a storm's development farther offshore, can radically change the forecast for a given area. Nearly one year later, there was an encore. With very cold air in place over the eastern third of the United States, a strong upper-air low moved southeastward into Texas and the Gulf coastal region, where snowstorms are born. There was consistency in the model guidance four days out; the Gulf storm would redevelop off of the South Carolina coastline and blossom into a winter storm, a nor'easter of historic proportions.

The potential for an East Coast blizzard was unmistakable. Like the year before, social media was on fire with commentary and, at times, dire messages. There were forecast challenges, and these concerns were made clear in the forty-eight hours prior to the storm's expected arrival. The dense arctic air could serve to suppress the axis of very heavy snow south, increasing the threat to Philadelphia and points south. The Arctic and North Atlantic oscillation were negative, a robust hint for a major East Coast storm.

However, some forecast models diverged on whether all of New Jersey would receive a crippling blizzard. Winter storm watches were issued Thursday, January 21, statewide. Initially, only light snowfall amounts were forecast for northwestern New Jersey and southern New York State. As of Friday morning, a general six to twelve inches was forecast for New York City into Northern New Jersey. By noon Friday, the entire scenario had

A cleared path reveals thirty-inch snow depth in Warren County, Blizzard of December 26–27, 2010. *Courtesy of Dave Dabour, New Jersey Weather Observers.*

changed. The morning run of computer models had shifted the heavy snow northward, covering the entire Tri-State area into southern New York State and all of New England.

A blizzard warning was issued Friday for all of New Jersey except the southeast coast, with heavy snow expanding northward at 2- to 3-inch-per-hour snowfall rates. The rapid intensification of the coastal cyclone, as forecast by model guidance, had taken place. Gale warnings were issued for the coastal plain with extensive coastal flooding and substantial beach erosion, most notably in Ocean City, Wildwood and Cape May, with a 9.4-foot storm surge inundating Cape May. Snow overspread Southern New Jersey by evening, reaching Newark and Northern New Jersey by 10:00 p.m. By midnight, heavy snow was in progress statewide, and by daybreak Saturday, wind gusts of over sixty miles per hour were lashing the Jersey shore. The blizzard buried all twenty-one New Jersey counties, with lesser amounts of snow covering the southeast coast, and walloped the entire metropolitan area through late Saturday night.

In Central Park, 27.6 inches of snow accumulated; 28.1 inches were recorded in Newark, 22 inches at Manalapan, 35 inches in Pine Hills Township and 33 inches at Morris Plains, nearly three feet of snow. Cars were buried, homes were enshrouded with snow, which drifted to five to eight feet deep, making many rural and county roads unidentifiable. Eight people perished and one hundred thousand people in New Jersey lost power Saturday during the height of the blizzard.

United Airlines suspended all flights from Newark Liberty Airport Saturday afternoon, and over ten thousand flights were canceled along the Eastern Seaboard. The blizzard was categorized as "crippling" on the NESIS (Northeast Snowfall Impact Scale), one of the top five blizzards experienced in New Jersey since the Blizzard of 1888.

On January 22–24, 2016, a large portion of the eastern United States was struck by a historic nor'easter. All the major metropolitan areas on the Eastern Seaboard were affected, including Washington, D.C.; Baltimore; Philadelphia; New York City; and Boston. The storm brought heavy snow, gale-force winds and blizzard conditions to many areas, while near-hurricane-force winds resulted in extreme coastal flooding. Commerce and travel were severely impacted for several days.

The deck of New Jersey resident Nick Stefano after the blizzard of January 22–23, 2016, dropped a foot of snow in Sparta. *Nick Stefano, New Jersey Weather Observers.*

A snowplow battles blizzard conditions and three-foot snowdrifts during the Blizzard of January 22–23, 2016. *Courtesy of Dave Dabour, New Jersey Weather Observers.*

The storm's total snow accumulations exceeded two feet across parts of West Virginia, Virginia, Maryland, Pennsylvania, New Jersey and New York, with snow accumulations reaching up to three feet in some areas. Snowfall records were established at many locations. The storm was later rated as a category 4, or "crippling," on NOAA's Northeast Snowfall Impact Scale.

Some select snowfall reports and records from official observing locations, storm totals: Somerset, Pennsylvania, and Allentown, Pennsylvania, 35.8 inches 31.9 inches (also 30.2 inches for calendar day record); JFK/Kennedy International Airport, New York, 30.6 inches (also 30.3 inches for a calendar day record); Harrisburg, Pennsylvania, 30.2 inches (also 26.4 inches for a calendar day record); IAD/Dulles International Airport, Virginia, 29.3 inches; BWI/Baltimore-Washington Airport, Maryland, 29.2 inches (also 25.5 inches for a calendar day record); LGA/LaGuardia Airport, New York 28.2 inches (also 27.9 inches for a calendar day record New York City/Central Park, New York City, New York, 27.5 inches (also 26.6 inches for a calendar day record); EWR/Newark International Airport, New Jersey, 24.5 inches; Philadelphia, Pennsylvania, 22.4 inches (also 19.4 inches for a calendar day record).

Meteorologist Dan Zarrow

Dan Zarrow is chief meteorologist for Townsquare Media, New Jersey 101.5 FM radio. On a typical day, Zarrow leaves home at 1:30 a.m., arrives at the station, analyzes the latest model data and develops forecasts for twelve different New Jersey radio stations between 4:30 a.m. and 7:00 p.m. Known as "Weatherman Dan" in his hometown of Jackson, New Jersey, due to his passion for weather, Zarrow has a career as a television meteorologist that has taken him to such diverse locales as Oklahoma and Central New York.

A graduate of Cornell University, where he earned a bachelor's degree in atmospheric science, he has done research and programming work for the Northeast Regional Climate Center and New Jersey State Climate Office under New Jersey State Climatologist David Robinson. In 2014, Zarrow replaced the iconic and nationally acclaimed chief meteorologist Alan Kasper, who retired after a forty-year television and radio forecasting career. Zarrow resides in Clark, New Jersey, with his wife, Amy, and their four young sons. A devoted husband, father and community member, he is the voice of New Jersey forecasting.

> *The Blizzard of 2016 was one of those weather events that can be rightfully classified as "a big one," one that shuts down New Jersey for days on end, sending residents scrambling to stock up on shovels, salt and French toast supplies. It's one that meteorologists both delight in and dread because of the forecasting nightmare and potential loss of life.*
>
> *Social media started buzzing about the potential for a powerful nor'easter about a week before. As is true of most coastal storms, forecast uncertainty surrounding the storm track was incredibly high. Early snowfall estimates literally ranged from nothing to two feet.*
>
> *The storm presented two special challenges:*
>
> *1. In late January 2015, a very similar coastal storm developed with a very similar setup. It was supposed to be the first "big storm" I covered as chief meteorologist for New Jersey 101.5. Forecast snowfall: eighteen to twenty-four inches for much of New Jersey. Actual snowfall: zero to nine inches, as the storm swung fifty miles farther east than expected. The wounds of "the blizzard that wasn't" were still fresh one year later, as we explored new ways to effectively communicate an uncertain forecast in the social media age.*

2. The weekend timing. The brunt of the Blizzard of 2016 struck New Jersey from late Friday night into Saturday. On the one hand, there are fewer people "out and about" on nights and weekends, which lessens the danger of icy roads. But it still had to be an "all-hands-on-deck" situation for the radio station, from the newsroom to the digital desk to backup hosts to the engineering team. (At the time, New Jersey 101.5 was one of very few radio stations that had live, in-studio personalities on-air 24/7.) Behind the scenes, almost twenty people camped at the radio station, sleeping on the floor for the weekend.

Given the duration and severity of the storm, I left my wife, infant son and dog safe at home and took up residence on an air mattress in the NJ 101.5 promotions office from Thursday to Sunday.

In the radio world, coverage during a snowstorm is relatively calm—there are flurries of activity, alternating with lots of waiting. We watched the snow fall and listened to the wind howl, as phone calls and photos came in from listeners across the state. I remember the most dramatic time was the approach of high tide, as coastal waterways crested over major flood stage across New Jersey's southern shore. (This was, of course, the storm

Whiteout conditions combat a snowplow in Warren County during the height of the great blizzard of January 2016. *Courtesy of Dave Dabour, New Jersey Weather Observers.*

where Governor Christie asked if he should "grab a mop" to help with flood cleanup efforts.)

Covering the Blizzard of 2016 was truly a team effort, both on the air and beyond. Once the storm started to wind down Saturday afternoon, we managed to enjoy a hot meal after finding the one and only Chinese restaurant in town that was actually open and delivering. On Sunday, our radio team transitioned to a shoveling team to clear our cars and the parking lot to finally go home.

I often remind my audience of the perils of winter season forecasting that "all it takes is one big snowstorm to make a season memorable." The winter season of 2015–16 was clearly a memorable one because of the historic snowfall and dramatic flooding of the Blizzard of 2016.

New Jersey State Climatologist David A. Robinson

The fragile barrier islands are the most vulnerable land areas in New Jersey. Charming, resplendent with nature and the allure of the ocean, the islands tell their own story. Island residents understand they are at risk, and all could be lost with the next nor'easter or hurricane.

Dr. David A. Robinson is a distinguished professor and New Jersey state climatologist who works with the Department of Geography and New Jersey Agricultural Experiment Station at Rutgers University. He's a scientist whose research interests run the spatial gamut, from global to local, with the underlying theme being the development of a better understanding of the climate system.

He has been the face of New Jersey climatological and statistical information since he was appointed state climatologist in 1991. From Cape May Courthouse to High Point, he has provided public information on a multitude of winter storms that have visited the Garden State, none more memorable than the great Blizzard of January 22–23, 2016.

Despite New Jersey's winters continuing to warm, there have been a goodly number of major snowstorms in recent decades. One of the grandest was the January 2016 blizzard. Like many such storms, it delivered a multitude of conditions across the state. This included hours of blizzard and whiteout conditions in some locations, major coastal flooding and snowfall totals ranging upwards of thirty inches.

Heavy snow overspreads Warren County during the blizzard of January 22–23, 2016. *Courtesy of Dave Dabour, New Jersey Weather Observers.*

New Jersey was hit square on, with the most snow falling from Warren and Hunterdon Counties eastward to Monmouth, Union and Hudson Counties. Meanwhile, southern Cape May County saw just seven inches (sleet and rain mixed in for a time), and northwest Sussex County saw six inches, just missing the heaviest snow bands.

As the New Jersey state climatologist and a snow scientist, I saw the storm in terms of its dynamics, statistics and impacts and with a historical perspective. Fortunately, the worst of the storm came on Saturday, January 23, thus sparing the region of roadway gridlock or a multitude of individuals stranded at school or work.

Still, it was a dangerous storm, and tragically, eight New Jersey residents lost their lives in storm-related incidents. Coastal flood damage was significant, particularly in the south. Despite winds gusting over sixty miles per hour along the coast and over forty miles per hour at exposed inland locations, power outages were fortunately not widespread. This likely due to the mostly dry windblown snow failing to cling to branches and powerlines.

On a personal note, as a snow enthusiast, I was outside experiencing the challenging conditions on multiple occasions, several times on cross-country skis and others with shovel in hand. At its greatest depth in my mostly wind-sheltered backyard, an impressive twenty-two and a half inches of the white stuff was measured.

This certainly was not an event to take in by simply looking out the window, nor wearing my state climatologist's hat and conversing with several dozen reporters during the storm and over the days surrounding it. As with many snowstorms I've experienced in my home state since the 1960s, the January 2016 storm will forever be fondly ingrained in my memory.

The Nesenohns

Alan and Gerri Nesenohn had their home built on Sea Isle City in 2007 and relocated from their home on the mainland, where they had lived since 1991. Alan was a technical school culinary arts instructor for years, mentoring students in the various aspects of food preparation. One of his students received the prestigious James Beard Foundation Award, given in recognition of chefs, restaurateurs, authors and journalists in the food industry.

Gerri Nesenohn

My husband, Alan, had just gotten back from the hospital. I had planned a trip to Florida to visit friends, yet with my husband being home and warnings for a big nor'easter, I had to cancel my flight. I had to provide verification of the expected storm to the airline before I was refunded. There was flooding at high tide, and we lost a lot of sand. Most residents did not have grass on the island; they had stone, and the water rushing down the streets lifted the stone from many homes into the street.

Alan Nesenohn

When I woke up, there was layers of ice right up to our doorstep. We lived on Seventy-Ninth Street. We lived in Souderton, Pennsylvania, and had a summer house in Sea Isle City since 1991. We had a house built in Sea Isle in 2007 and moved there. [There were] *eight inches of ice and slush from the snow that fell earlier, which was followed by heavy sleet and rain. There were ten- to twelve-foot waves that carried the stones into the street. The ocean breached the dunes; there was not much beach left. You couldn't*

A rural Warren County snowscape the morning after the great blizzard of 2016. *Courtesy of Dave Dabour, New Jersey Weather Observers.*

Snow fun in the blizzard. A solitary figure in the heavy snow during the blizzard of 2016. *Courtesy of Dave Dabour, New Jersey Weather Observers.*

A house buried in eight-foot drifts after the blizzard of January 2016 in southern Sussex County. *Courtesy of Dave Dabour, New Jersey Weather Observers.*

tell high tide from low tide there was so much water. The water reached our first step.

Fortunately, it was winter, and there were not many people on the island; it was barren. Our block goes from the ocean to the bay, and the entire area was inundated. The streets were flooded the tidal flooding was so severe.

16

THE "SNOWVEMBER" SNOWSTORM OF 2018

Thousands of New Jersey motorists were stranded on local roads and highways, as well as the New Jersey Turnpike and the Garden State Parkway. The police responded to 1,000 crashes and assisted 1,900 struggling motorists, many taking over four hours to get home on commutes that would normally take an hour. New Jerseyans, from the school crossing guards to Governor Phil Murphy, were shell-shocked.

The first snow of the season was in the forecast, yet forecasters tamped down expectations—a coating of an inch or two before changing to rain. This was not a nor'easter, just a late fall rainmaker with a tiny taste of the winter to come. No alarm bells were sounded—it was only mid-November, nothing to worry about!

New Jersey is the fourth-smallest state in the nation. However, it has the tenth-largest state economy, which reached $586.8 billion in 2023. That translates to busy residents! From those working in large industries to the corner convenience store clerks, New Jerseyans are in a state of perpetual motion. No state is more reliant on accurate weather forecasts as the day begins, particularly when a winter storm is expected.

Meteorologists are aided by sophisticated model guidance, which interprets vast amounts of complex data and issues accurate five-day forecasts 90 percent of the time and accurate next-day forecasts 98 percent of the time. However, sometimes Mother Nature has something else in mind. Thursday, November 15, 2018, was one of those times. From 12:00 p.m. through the nightmare evening commute, residents from Monmouth and Middlesex Counties northward through all of northern and northwestern New Jersey experienced what can only be described as pure hell.

Governor Phil Murphy held a storm briefing the following morning and took full responsibility for the nightmare on the roads. "I can understand and appreciate completely what I am hearing, and our team is hearing from commuters," one of whom was his predecessor, former Governor Chris Christie, who spent five hours and forty minutes driving from Piscataway to Mendham. The irate Christie did what dozens of motorists did that afternoon: he called New Jersey 101.5 to complain. "I could tell you this, I've never ridden in anything like this," he said. "Dead stop," he told New Jersey 101.5 hosts Jeff Deminiski and Bill Doyle. New Jerseyans were unprepared; November snowstorms are a rarity. This was a nightmare.

FRANK INFANTINO

A former Staten Island resident who moved his young family to Bound Brook in 1998 and then to Philipsburg in far-western New Jersey in 2005, Frank Infantino was no stranger to snowstorms. His commute to Turtle Inc. in Bridgewater took fifty minutes. An inside salesperson and former department manager, Frank is a people person, a dedicated Yankees and Miami Dolphins fan and a great storyteller. He is patiently waiting for the Yankee's twenty-eighth World Championship title.

> *As I drove in that morning it was cloudy and cold, more January-like than November. There were already patches of ice on the side roads leading onto Route 78. I had the radio on, light snow and sleet was in the forecast, then rain. It was a Thursday, almost the weekend. It was a bye week for the Dolphins, and I would have to wait until Thanksgiving weekend, Sunday, for the next game against the Colts.*
>
> *The morning was busy; I had to work on a very large project with a lot of wire, panelboards and enclosures. I usually keep two snow scrapers and a small shovel in my car during the winter. Usually, when they forecast snow, that's all you hear about on the TV and radio. I went out to grab a sandwich for lunch, and it sure felt like snow. We had what they call the "snow sky."*
>
> *On my way back to the office, it started to snow. Steady snow by 1:00 p.m., and then the ground and roads were already covered. The snow was*

wet, and as the day went on, it was getting serious. I went out to clear off my car, already two inches had fallen by 3:00 p.m. I took one call after another from customers with horror tales about road conditions. Now there were advisories, and four to six inches were expected. There wasn't one of my coworkers who wasn't in shock and worried about the drive home, as I was. I left at 5:00 p.m.; the snow was pouring down, four-plus inches on the ground and not a snowplow in sight. It took me a half hour to finally get onto 78 West.

I was trapped on Route 78 West between exits because the trucks couldn't make it over the hill and kept backsliding. There was no way out, bumper to bumper. Inching along. Then the snow began changing to sleet and ice. I had no cellphone to call anyone. I ended up getting home at 11:30 p.m. that night, six and a half hours of agony. Then back to work the following morning with most of the roads still clogged with snow and ice.

Ed Miro

Ed Miro, known as "Big Ed," is an outside sales representative for Thea Enterprises, a manufacturer's representative agency for electrical construction and lighting supplies. Ed has been with Thea for over thirty years. A gregarious man and recognized salesperson (he could sell sand at the beach), Ed, with his wife, Beth, has two adult children and lives in Colts Neck, New Jersey.

It was a wintry morning, cloudy and cold, I had to be at the Plainfield branch location of Turtle & Hughes, where I was doing a pre-Thanksgiving counter day demo from 8:00 a.m. to 12:00 p.m. I had coordinated the event with Herman Schwalbach, the counter manager. We had a big two-table display of Klein tools, many of which I demonstrated for the contractors, with a grand prize raffle, giveaways and a hot Thanksgiving turkey lunch.

As expected, the counter was crowded with electrical contractors and industrial customers, so we had a large turnout. One of the customers was from the North Plainfield Board of Education, and he told me, right before noon, that it was snowing in parts of Central Jersey and the roads were already covered. I lived near the shore, where it usually rains, so I wasn't worried. Shortly after noon, I began to close down the counter [for the] *day and pack up my car to drive home, as I was done working for the day. I saw snow in the air, and before long, it was coming down hard.*

I stayed at the branch to complete some work and talk with a few more customers. I left for home at 1:30 p.m., and by then, it was snowing heavily. Normally, it is a fifteen-minute drive from the Turtle Plainfield store to Exit 135 on the Parkway. It was the winter's first snow, and it caught drivers off guard. This was a real snowstorm. It took me two solid hours to make it to the parkway. The traffic wasn't moving a bit; schools were letting out, and it had become pure havoc.

The normal hour drive home to Colts Neck took me four hours. There were many accidents, some nasty. There wasn't a sign of a snowplow, and before I got home, it was now freezing rain on top of four inches of slush. I've had easier drives through blizzards. I was glad to get home and that [the next day] *was Friday.*

17

A CENTURY OF NEW JERSEY SHORE HISTORY

CAMP OSBORN

Beachfront living had its origins on New Jersey's barrier islands, with their many cottage and bungalow communities that took root at the water's edge over a century ago. By the late nineteenth century, small towns began to dot the smaller barrier island that is also home to Island Beach State Park.

Seaside Heights was incorporated as a borough in 2013. The Borough of Lavallette was formally incorporated in late December 1887, but Lavallette, as a named location, is almost ten years older. In February 1878, the directors of the Barnegat Land Improvement Company filed a plot plan with Ocean County, designating the tract they purchased from Michael W. Ortley as "Lavallette City by the Sea."

Ortley Beach was settled by Michael Ortley in 1812, and beachfront development in the area did not begin until over a century later. The towns of Point Pleasant and the historic Jenkinson's Pavilion, Bay Head, Mantoloking and Brick are each unique in their individuality and charm.

Grand hotels, such as the Manhasset Hotel, built at Seaside Park at the turn of the century, attracted visitors from Philadelphia and New York City. A growing fishing and tourism industry grew the local economies and provided jobs, even during the Great Depression. The largest of the barrier islands, Long Beach Island, is nineteen miles long and has long been the centerpiece of the Jersey Shore economy.

There is a special story on the island, a generational story of children, moms and dads, grandparents, beach chairs, surf rods ready for bluefish, barbecues, fireworks, laughter, cold Coors Light, rusty old bicycles, birthday parties, little girls playing hula hoops. When there's a birthday, expect a knock on the door and a piece of cake. Neighbors say goodbye in October, only to reunite with hugs in May. Newborns are christened, elders sit in front of their bungalows to watch yet another sunset, one of thousands they have seen set over the bay side of Camp Osborn.

Camp Osborn saw its beginnings during World War I as a patchwork of tents and wood shacks with cold showers for beach fishermen. Originally located on the ocean side of the island, by the 1920s, what eventually became Marion, Cummins, Shell and Elder Streets along Route 35 had become a summer gathering place for friends and families. Children played on the "Boulevard." Women carried parasols, and men wore one-piece swimsuits, common to the era. Neighbors and families networked into a community. Camp Osborn was born.

The Camp was all about stories; it became generational, with the shacks being built into bungalows by World War II, charming cottages built close together, summer homes that were passed down to children and grandchildren, along with the memories and recitals of summers past. The bungalow community had also survived powerful nor'easters, tropical storms, hurricanes and crippling blizzards and violent thunderstorms.

The powerful nor'easter of January 27, 1933, which ripped through the barrier islands, resulted in major coastal and backbar flooding, accompanied by gales and heavy rain and snow. The category 5 September 1938 hurricane that grazed the coastline resulted in significant damage and extensive flooding on the barrier islands. Back-to-back Hurricanes Carol and Hazel in 1954, the Great Atlantic Storm of 1962, which all but obliterated Long Beach Island, the massive nor'easter of December 11–12, 1992, the deadly "Storm of the Century" in March 1993, the blizzards of February 1978 and January 2003 and the Christmas weekend Blizzard of 2010—with the exception of flooded streets and a few missing roofs, Camp Osborn withstood them all.

Until Sandy.

Stan Kosinski

Stan Kosinski's history in Camp Osborn began in 1995, when his dad bought a hot dog for himself and saw an advertisement for a home for sale at the shore for $30,000. He decided to look into it and bought the home on the spot. A native of Elizabeth, Stan spent two decades in the timeless beach community before it was consumed by fire during Superstorm Sandy in 2012. Kosinski and his family are a part of the history of the Camp, and he has his share of "Camp stories."

> *I spoke with an old fisherman on the beach who told me about rum running and the Camp. It was the farthest southern development at the time; they would dump the barrels and let them wash up on the beach and then hide them in the bungalows to be picked up. The chief of police would get his barrel and look the other way. I didn't know who he was, the old fisherman, or if there was any truth to it. This was in 1995 when he told me this story.*

Dan Smith

One of the World War II generation Camp Osborn residents, Dan Smith lived in a bungalow on East Marion Street. Many of the original cottages were stick built by hand, using a hammer, nails and wood—along with a few cold beers. Home by home, the personality of this charming beach community took shape from the 1920s through World War II.

Wayne Smith recalls his father, Henry, and his uncle Dan, who, through hard work and a bit of creativity, played a role in the storied history of Camp Osborn.

> *My father, Henry, and his brother, my uncle Dan, and two other guys somehow got an old tool shed located on the ground of Metuchen High School. They loaded it on a truck and drove it to the Camp in the dark because it was an oversize load. It had six bunks in it (they were three feet high), a hutch for the dishes and a kitchen and toilet closet.*
>
> *The shower was outside. At some point, they acquired, by rental agreement, a vacant lot between them and the next house east. The places were tiny, but we all had great times there. This was around 1939–40, before the war. There were so many memories in that little bungalow, like the night me and a bunch of college buddies were awakened by a state trooper at 2:00 a.m. to evacuate during a hurricane in September 1967. When we came back in the morning, the waves were running, and we decided to go surfing—not a good idea!*

In the very early years, there were cold-water outdoor showers and a single telephone booth. Connie and Piela purchased bungalows in the early 1940s and repaired them. The Taylor, Cavanna and Mavus families always rented bungalows and eventually bought into the Camp when the cottages were sold in the early 1960s. Families rented every summer, and friendships and summer traditions grew over the years. The Camp expanded into four sand streets, with houses on the median of Route 35 and the bay side.

The Thunderbird Motel, a landmark built next to Camp Osborn, became a focal point for surf fishermen due to the large channel behind the motel, a deep pool of ocean water that attracted schools of baitfish, a feeding ground for big bluefish and striped bass during the spring and fall migrations.

March 12–14, 1993

By Friday, March 12, 1993, shore residents along all 141 miles of the Jersey mainland beaches and the barrier islands were frantically making preparations for what was forecast to be one of the most powerful coastal storms to ever impact the Eastern Seaboard. It earned the epithet "Storm of the Century."

The approach of this extraordinary coastal storm had already laid a trail of death, destruction and billions of dollars in damages. Half a foot of snow had fallen in the Florida Panhandle, and all sixty-seven counties in Alabama reported snow cover. Millions who made storm preparations, fueled by intense media coverage, were driven by fear as well as necessity.

The 1993 storm did not quite rival the destruction wrought upon New Jersey during the 1944 hurricane or the Great Atlantic Storm of 1962 in terms of damage, destruction, loss of life or geographical impact. The forty-eight-hour winter hurricane spawned deadly tornadoes and hurricane-force wind gusts. A gust of 109 miles per hour was recorded at the Dry Tortugas in Florida, along with more than $3 million worth of property damage in twenty-two states. Thirty-foot swells in the Atlantic and the Gulf of Mexico prompted the rescue of over two hundred people from foundering boats.

Temperatures were in the middle to upper thirties at the New Jersey immediate coast, yet the growing northeast gales kept enough cold air in place at mid-levels of the atmosphere to allow heavy snow to surge northward, arriving in Cape May before midnight. The snow powered northward through Atlantic City and Long Beach Island, reaching Camp Osborn in

Northern Ocean County by early Saturday. With the exception of several year-round Camp residents, the bungalow community was barren.

Five inches of snow had accumulated by 8:00 a.m. Saturday, with 35- to 40-mile-per-hour sustained winds. The water washed over the dunes at the Camp; a rare occurrence witnessed only a handful of times in the near eighty-year history of Camp Osborn. Only 1,300 feet separated the ocean from the bay, a three minutes' walk, the narrowest point on the barrier island, making it extremely vulnerable to storm surge, as would be witnessed nineteen years later during Sandy.

By mid-morning, the raging nor'easter, easily the equivalent of a category 1 hurricane, was marching northward, paralleling the New Jersey coastline, a track that allowed mid-level atmospheric warming, rapidly changing the snow to sleet and then to plain rain at the beaches. Narrow streets on the island were compromised as a result of ocean and back bay flooding.

Because the storm was accurately forecast up to five days in advance, preparations were largely complete before the great storm, one of the deadliest and costliest winter storms in the nation's history. The Camp Osborn bungalows were pounded with seventy-mile-per-hour wind gusts, yet they bore the storm well, as they had the great 1944 hurricane, the 1962 Ash Wednesday storm and the blizzards of 1978, 1996, 2003, 2006 and 2010. They were battered but not broken.

Until Sandy.

18
NOT HISTORIC, YET NOTABLE WINTER STORMS

The sixteen winter storms outlined in this book and the stories of those who lived through them were historic because of their societal impacts—destruction, economic cost, life-altering outcomes, loss of life. Many had stories of survival; some had remembrances of joy and family, such as the white Thanksgiving of 1989, a rarity, a snowstorm in November. "Over the River and Through the Woods," the poem by Lydia Maria Child, came to life and made so many families happy on this day when families are celebrated. It was the biggest Thanksgiving snowstorm (there haven't been many) on record.

There was the shocking surprise of November 15, 2018—not that a half foot of snow was historic, but there were five unexpected hours of heavy snow. It was not in the forecast and left thousands of furious New Jersey commuters trapped on county roads, the New Jersey Turnpike and Garden State Parkway for up to six hours. Snowbound motorists called NJ 101.5 in outrage, as normally forty-five-minute commutes took up to four hours. In terms of misery and aggravation, the "Snowvember" storm may have rivaled the Blizzard of '96.

There were other storms that merit recognition. They were not historic, yet they are firmly secure in the memories of New Jerseyans, both young and old, as New Jersey was at winter's mercy. We will turn back the pages of time and revisit twelve of these winter storms.

THE BLIZZARD OF JANUARY 24–25, 1905

Adding additional misery to an already harsh winter, a moisture-laden low-pressure area moved eastward from the Ohio valley and redeveloped into a paralyzing winter storm that moved along the New Jersey coastline, bringing wind-driven heavy snow to New Jersey.

Residents of northwestern New Jersey were forewarned of the pending snowstorm by means of telephone and telegraph. Not everyone had a telephone at the turn of the century. The crippling blizzard had wind gusts of over fifty miles per hour driving the powder into huge drifts. Travel was near impossible.

Over a two-day period, eighteen to twenty-four inches of snow accumulated, with drifts reaching ten feet, leaving thousands housebound. It shut down towns and cities and incapacitated businesses. It was the most severe winter storm since the Blizzard of March 12–14, 1888.

Two horsedrawn people navigate snow-covered roads after a nor'easter brings a massive snowstorm to Sussex County. *Courtesy of the Sussex County Historical Society.*

In wooded Sussex Boro, a man has a tranquil moment after thirty inches of snow bury northwestern New Jersey. *Courtesy of Wayne T. McCabe, president of the Sussex County, Historical Society and National Park Service, qualified architectural historian.*

Residents of Sussex Boro were homebound as drifts from the January 24–25, 1905 blizzard smothered their homes. *Courtesy of Wayne T. McCabe, president of the Sussex County Historical Society and National Park Service, qualified architectural historian.*

Top: A man stands in front of a confectionery while, across the street, a man shovels snow in the aftermath of the blizzard of January 1905, Newton, New Jersey. *Courtesy of Wayne T. McCabe, president of the Sussex County, Historical Society and National Park Service, qualified architectural historian.*

Bottom: The crippling blizzard of January 23, 1905, turned downtown Newton, New Jersey, into a ghost town. *Courtesy of the Sussex County Historical Society.*

The face of winter. A man shovels snow to clear a path for pedestrians in Newton, New Jersey, after the Blizzard of January 24–25, 1905. *Courtesy of Wayne T. McCabe, president of the Sussex County, Historical Society and National Park Service, qualified architectural historian.*

People in the center of a snow-covered street in tiny Sussex Boro, New Jersey, after the great Blizzard of 1905 buries the community. *Courtesy of Wayne T. McCabe, president of the Sussex County, Historical Society and National Park Service, qualified architectural historian.*

A person struggles in the snow outside of the Newton Mission after thirty-inch snowfall buries the town, January 24–25, 1905. *Courtesy of Wayne T. McCabe, president of the Sussex County Historical Society and National Park Service, qualified architectural historian.*

A municipal building in Newton, New Jersey, serves as a backdrop after the January 5, 1905 blizzard buries New Jersey. *Courtesy of the Sussex County Historical Society.*

A man is dwarfed by the snow in front of Whitman's chocolates and confections, Sussex Boro, New Jersey, Blizzard of January 1905. *Courtesy of Wayne T. McCabe, president of the Sussex County, Historical Society and National Park Service, qualified architectural historian.*

A man takes a break from shoveling against the backdrop of ten feet of snow Newton, New Jersey, 1905. *Courtesy of Wayne T. McCabe, president of the Sussex County, Historical Society and National Park Service, qualified architectural historian.*

The Snow Hurricane, March 1, 1914

Twenty-four to thirty inches of snow hammered the New Jersey landscape with hurricane-force wind gusts, the result of an extremely intense coastal storm that ground much of New Jersey to a standstill.

New Jersey Governor James F. Fielder, while en route to New York City, was one of hundreds of Pennsylvania Railroad commuters stranded at Elberon. Asbury Park was particularly hard hit, leading the *Asbury Park Press* to proclaim: "Asbury Park Cut Off from the World."

December 11–12, 1960

In an unprecedented early season snowstorm, twenty to twenty-eight inches of snow buried Central and Northern New Jersey, as a deep low-pressure center moved slowly up the Eastern Seaboard, east of the New Jersey coastline, barely a month after John F. Kennedy's election as the thirty-fifth president. This was the first of three high-impact snowstorms in New Jersey during the winter of 1960–61.

February 3–4, 1961

The second snowfall of 20 or more inches during the winter of 1960–61 walloped New Jersey on February 3–4, 1961. The third in a series of East Coast snowstorms rode the 40–70 benchmark, maintaining a supply of cold air combined with driving, heavy snow. Newark Airport recorded 22.6 inches of snow. Over two feet blanketed much of Northern New Jersey during a winter that delivered over 60 to 70 inches of snowfall to much of the state.

The Christmas Eve Snowstorm of 1966

New Jersey saw quite a white Christmas as a strong low-pressure system brought bands of moderate to heavy snow to much of the state, with mixed precipitation falling along the southeast coast; 7 inches fell at Newark, and 7.1 inches fell at Central Park. The baby blizzard made easy sledding for Santa, as Christmas Day arrived in a mantle of white. Many older New Jerseyans fondly remember this special snowstorm.

THE NEW YEAR'S DAY SNOWSTORM, JANUARY 1, 1971

A rare New Year's Day snowstorm, well forecasted by the National Weather Bureau and television and radio outlets, delivered 6 inches of snow to Newark and 6.4 inches to Central Park in Manhattan. It was a fast-moving coastal storm, and its snow accumulated in a four-hour period from 3:00 to 7:00 a.m. and exited quickly. New Jersey has not seen a New Years' Day snowstorm since.

DECEMBER 26, 1976

Despite repeated assurances of no white Christmas for New Jersey, this storm, originally forecast to track inland and bring rain, surprised forecasters by developing just east of the coastline, locking in the marginally cold air mass that was in place. Rain and sleet developed late Christmas night and changed to snow quickly north of Trenton, bringing a three- to seven-inch snowfall to Central and North New Jersey.

THE BLIZZARD OF 2000

As New Year's Day approached, a small, intense nor'easter developed off of the Virginia coastline and raced northeastward. Having characteristics of a winter version of a tropical storm, it snowed relentlessly for ten hours. Hamburg received twenty-four inches, Sussex got twenty-five inches and twelve- to sixteen-inch accumulations were widespread through much of New Jersey, with some lower totals in far southern areas of the state. The swirling snowstorm fell on the Saturday of the holiday weekend, with the millennium ending on a wintry note.

THE VALENTINE'S DAY SLEET STORM OF 2007

Forecast models were in full agreement that a moisture-laden winter storm would arrive in New Jersey on February 14. Two to four inches of snow were forecast for much of New Jersey as a low-pressure area hugged the coast. Snow developed, leaving a quick one to two inches, and then changed to an all-day-and-night, wind-driven, stinging sleet. Five inches of sleet accumulated, a winter rarity in the Garden State. A thin layer of warmer air just above the surface changed falling snow to stinging ice pellets, leaving New Jersey, once again, at winter's mercy.

THE HALLOWEEN SNOWSTORM, OCTOBER 29, 2011

For days, forecasters were in agreement that the Saturday before Halloween would herald the arrival of a coastal storm for New Jersey. A supply of unseasonably cold air for late October raised the specter of heavy, accumulating wet snow for the interior northeast.

By Friday night, the National Weather Service and Tri-State area forecast venues warned of a historic early season snowfall over the Pocono region of northeastern Pennsylvania, extending into the northwest corner of New Jersey, southern New York and the interior Connecticut, with power outages and downed trees likely. Rain developed in New Jersey during the pre-dawn hours Saturday and rapidly mixed with and changed to heavy, wet snow over central and northern areas of the state by mid-morning. The changeover to snow closer to the coast, from Somerville to Newark and New York City, caught forecasters off guard. It was a historic surprise.

A winter storm warning, the first ever issued in October, was issued for Sussex, Morris, Warren and Hunterdon Counties, with winter weather advisories hoisted for much of Central and Northern New Jersey. The impacts were staggering. Snow fell all afternoon, mixed with some rain at times, and additional light snow fell into Sunday morning.

Rain occurred over Southern New Jersey. Winds gusted up to fifty-four miles per hour at Tuckerton Shores in Ocean County; 5.2 inches accumulated in Newark, with 17 inches recorded at Lake Hopatcong, 16 inches at Sparta and a measurement of 19 inches at West Milford. An estimated 700,000 New Jersey residents lost power, including Governor Chris Christie's home in Mendham. Thousands of trees collapsed under the weight of the heavy snowfall.

Eight New Jerseyans lost their lives. Halloween, which fell on the Monday following the storm, was canceled in many communities. Governor Christie stated that the effects of the Halloween snowstorm were worse than those of Hurricane Irene the previous August. A winter storm for the record books!

THE FOUREASTER

There were four nor'easters during that period. The atmosphere was stretched to an extreme for nearly five weeks from late February to late March in both the Atlantic and Pacific.
—Meteorologist Joe Cioffi

The "Foureaster," March 2–22, 2018, provided a weather anomaly. Four nor'easters battered the Garden State, the first three winter storms occurring from March 2 to March 15, bringing heavy rain, significant accumulating snowfall over interior areas of the state, gale-force winds and significant beach erosion and damages along the coastline. An extraordinarily active pattern was in place, with repetitive phasing of the northern and southern jet streams spawning significant coastal storms, nor'easters, along the United States' Eastern Seaboard.

The fourth and strongest of these winter storms heralded the arrival of spring on March 20. With cold air in place over New Jersey, low pressure developed over the Rocky Mountain region and tracked east-southeastward. Once again, the northern and southern jet streams merged, and a powerful low-pressure area redeveloped off the Mid-Atlantic coastline, near the Maryland-Delaware coastline.

Unlike the previous storms, this system had a midwinter component, with significant snow accumulations forecast for Central and Northern New Jersey, right down to the shoreline. Early winter storm watches were upgraded to warnings Tuesday night. Heavy snow developed Wednesday morning, spreading south to north across New Jersey.

By Wednesday afternoon, the center of low pressure intensified dramatically, deepening to 988 millibars, a powerful nor'easter; 15.5 inches of snow accumulation were observed at Berkley and Lacey Townships, and widespread 8- to 12-inch accumulation was seen over Monmouth and Ocean Counties, 8.3 inches were reported at Newark Liberty Airport and 8.4 inches at Central Park. Snowfall rates of 3 inches per hour shut down northern parts of the state.

THE GROUNDHOG DAY NOR'EASTER

New Jersey has not witnessed a high-impact, major snow event since the near-Blizzard of January 31–February 3, 2021. This was an extraordinary winter storm; after departing New Jersey, it slowed to a crawl off the New England coastline, sending bands of light accumulating snow westward back into New Jersey and prolonging the storm. Some towns in coastal southern Cape May County received mixed precipitation, with under an inch of measured snowfall, and 4 to 12 inches fell over Central New Jersey. Most of northwestern New Jersey saw 16 to 24 inches, with an incredible 30 to 36 inches reported in a narrow area of Sussex County; 35.1 inches were recorded at Mount Arlington, 33.2 inches in Montague and 32 inches at Lake Hopatcong.

The storm system had its origins as a low-pressure area that brought heavy rain to Northern California and heavy snow to the mountain region. It then brought accumulating snow to the Midwest, driving southeastward and redeveloping off the North Carolina coastline into an intense nor'easter with its central pressure dropping to 990 millibars.

Light snow developed late Sunday afternoon over New Jersey, with rain falling over the southeastern part of the state. By midnight, four inches had accumulated at Linden. Snowfall rates on Tuesday morning averaged one to two inches per hour in banding, with four-inch-per-hour rates witnessed over interior Sussex and western Passaic Counties. Governor Phil Murphy declared a state of emergency on Sunday, January 31, at 7:00 p.m. in anticipation of the storm, banning all nonessential travel. Although the storm did not meet the full criteria for a blizzard, this long-duration, imperiling snowstorm not only proved the Groundhog Day forecast wrong, but it also joined a long list of dangerous winter storms to visit New Jersey.

MY STORY

Every winter storm is distinct, with its own story to tell. I've recalled sixteen historic winter storms in this book, having interviewed meteorologists, historians, businessmen and truck drivers, all with a story to tell. Some recalled the joy of a Thanksgiving snowstorm, while others' remembrances were those of fear, even terror, and survival. I have an account of my own to tell about the time I tried to outrace a blizzard that was bearing down on New Jersey.

My absorption in meteorology and frozen precipitation has enabled me to treat each approaching winter storm like the box score in a baseball game. When I was a boy growing up in Elizabeth's North End, even a manageable two- or three-inch snowfall had me fixated on my tiny transistor radio, the one my uncle Albert sent me from Japan as a gift, waiting for the 5:00 p.m. revised Weather Bureau (as it was called in the 1960s) forecast. Sometimes, I would turn the dial to CBS 880 to get the latest from meteorologist Gordon Barnes.

The second week of February is normally the snowiest week, statistically, of the winter season in New Jersey. In the winter of 2005–06, I had recorded ten and a half inches of snowfall in Linden, and by Tuesday, February 11, there was speculation that a significant winter storm would develop off the Southeast's coast by Friday. The NAM and GFS forecast models were conflicting, and the initial forecasts for the February 11–12 weekend kept the storm system mostly south of New Jersey. There was little mention of snow in most forecasts.

I remember the respected veteran meteorologist Dave Bowers, in his Wednesday evening Accu-Weather forecast, saying, "A storm system to our south may bring a small amount of snow to our area." As was my wont, I followed the forecast outlets intensely, and within twenty-four hours, the forecasts and computer models had changed radically, and now, a full-blown snowstorm, possibly reaching blizzard proportions, would impact New Jersey late Saturday into Sunday.

By Friday, forecast guidance was in agreement that phasing of the northern and southern jet stream, with cold air poised to interact with the developing storm, would bring gales and heavy snow to most of New Jersey. A winter storm watch was issued for Friday, February 10, and was quickly upgraded to a winter storm warning by nightfall and a blizzard warning by Saturday morning.

So, one would think Saturday was not a good day to go down the shore.

My wife worked for the Linden Fire Department. A Linden fireman, Brian Krakovsky, had an auto repair shop in Brick, which was a few miles from our bungalow at Camp Osborn, where we spent our summers. Our cottage was seven hundred feet from the water's edge.

Brian did our auto work for us, and we had to drop our Dodge Caravan off at his shop for brake work and a tune-up. My wife and I drove down the shore in separate vehicles after work on Wednesday, February 8, to drop the van off. Brian told us the van would be ready Saturday, which meant another trip to Toms River to pick up the van. We had to get it back; my wife needed it for work.

Friday night, at the kitchen table, we had a decision to make. Television and radio were saturated with blizzard coverage, with heavy snow due to arrive late Saturday afternoon—and even earlier to our south. She needed the van. We decided to get up early Saturday, drop off the van and get home ahead of the snowstorm.

The following morning, we were out the door at 10:30 a.m., not as early as we would have liked. With our eleven-year-old son, James, in tow, we got a coffee at Dunkin Donuts and headed south on the parkway. It was overcast and not particularly cold; it was thirty-seven degrees when we left. Colder air was on the way and would interact with the developing storm, resulting in a historic snowstorm. We were in a race.

By the time we passed the Garden State Arts Center, wet snowflakes were dancing in the air—just scattering, not enough to wet the ground. We arrived at the shop to pick up the van and pay the bill. After a round of fire department shop talk with Brian, we simply had to make a pit stop at our beach house. We had enough time. We were still ahead of the storm—or so we thought.

We drove separately to Camp Osborn, which was fifteen minutes away. The Camp was in winter hibernation, yet there were several longtime residents who lived there year-round. We opened the house, put on the TV for James and enjoyed a plastic cup of wine on the small back porch. Our neighbor, John Harvey, came outside, and we chatted. John was a Camp historian, and he always had a story to tell. We were enjoying ourselves; it was comfortable, and the wind was still. Then it started to snow.

I was somewhat surprised. In 2006, there was no smartphone to instantly check the radar. I thought, being at the immediate shoreline, at the onset, the precipitation would be rain. It was not uncommon for Linden to have ten inches of snow while the Camp gets a soaking. I was excited at the arrival of snow at the beach; I had witnessed only half a dozen snowflakes in the thirteen years since we closed on our bungalow in 1993.

I took my libation and walked the entirety of Camp Osborn, which consisted of four small streets. It was a tradition of mine; I would always walk the Camp before our family left for home on summer Sunday evenings. This time, it was enchanting. The gentle snow cast an entirely different ambience over the bungalows.

When I got back to the bungalow, my wife and I realized that the Camp had worked its magic; it was nearing 4:00 p.m. We took our belongings and left, driving separately home to Linden, a forty-nine-mile drive from Mantoloking.

By the time I stopped at Dunkin Donuts for a large coffee for the ride home, the wet snow had changed to rain. A bit disappointed at losing the snow, I crossed the Brielle drawbridge and picked up the parkway north.

Moderate rain continued until I reached the Garden State Arts Center, where the rain mixed with and changed to snow within minutes. This was not the gentle, light, wet snow we had at the beach house. The ground quickly whitened, and the visibility dropped to a tenth of a mile. A motorist skidded abruptly and cut me off. The shield of heavy precipitation had overtaken me, and the farther north I drove, the colder it got. I called my wife on my cell; she was well behind me, still in the rain. I warned her to prepare for a quick change to heavy snow.

As I passed Middletown and Laurence Harbor and approached the Driscoll Bridge, I was almost blinded by heavy snow. The Parkway was now covered in slush and snow; the road was untreated and very slippery. I called my wife twice, and there was no answer. She rarely answered the phone while driving, yet I was very worried. By the time I crossed the bridge, traffic was crawling. It took forty-five minutes to get to the Linden exit, normally a ten-minute drive from the bridge.

I was almost panicked with worry for my wife and son. There was an inch and a half on the ground, having fallen in less than an hour. As I backed in our driveway, my cell rang—it was my wife. She had pulled over at a rest stop but gave no words of reassurance. "Stop calling—it's snowing like hell!" she yelled.

I would never again try to outrace a blizzard.

BIBLIOGRAPHY

Allen, Craig (chief meteorologist). Interview about the Blizzard of January 7–8, 1996, commute to the studio, keeping the public informed on WCBS television and radio. September 2024.

———. "The Storm of the Century." WCBS. Aired March 12–14, 1993. (Radio and television storm coverage.)

Armstrong, Tim (meteorologist). "The 1993 Storm of the Century: Impacts, Loss of Life, Multiple States." NOAA, National Centers for Environmental Information. March 13, 2017. https://www.ncei.noaa.gov/.

Auciello, Justin. "57 Years Ago Today, Ash Wednesday Nor'easter Was Lashing NJ." WHYY NJ Radio. March 6, 2019. https://whyy.org/.

Bellangy, Harry, and Kathleen C. Wyatt. "The Great Atlantic Storm of March 5–9, 1962; Christmas Weekend Blizzard of 2010; the Great Blizzard of January 22–23, 2016." The Greater Cape May Historical Society, Sulke Archival Collection. 2024. https://www.capemayhistory.org.

Bichao, Sergio. "Even Christie Got Stuck in the Snow for Hours! He Calls NJ101.5 For Report." NJ101.5 FM. November 15, 2018. https://nj1015.com.

Black, Lauren M., and Paul Mickle. "Impacts, Trenton Train Station, National Guard Rescues, Economic Impacts." *The Trentonian*, January 8, 1996. https://www.trentonian.com.

Blanchard, Wayne. "Blizzard of February 11–12, 1983: People Stranded on the Garden State Parkway, New Jersey State Police and National Guard Rescue." Deadliest America Disasters and Large-Loss-of-Life Events. February 1983. http://www.usdeadlyevents.com.

———. "Dec 26–27, 1947, Winter Snow Storm, New England/20, PA/2, NJ/31, NY/17–70–77." Deadliest America Disasters and Large-Loss-of-Life Events. October 11, 2021. http://www.usdeadlyevents.com.

"Blizzard Hits New York City." *New York Post*, December 27, 1947. https://nypost.com.

"Blizzard of February 11–12, 2003." News 12, New Jersey. 2006. https://newjersey.news12.com.

"Blizzard of 2016, Weather Underground Coverage: Jonas, NESIS 4, Crippling." Weather Underground. 2024. https://weather.underground.com.

Boro of Lavallette "Boro of Lavalette History." 2025. https://www.lavallette.org.

Burt, Christopher C. "The Blizzard of 1888: America's Greatest Snow Disaster." Weather Underground. March 12, 2020. https://www.wunderground.com.

———. "Meteorology of Blizzard of 1888: Synoptic Charts and Observations." Weather Underground. March 12, 2020. https://www.wunderground.com.

"Camp Osborn." In *Boro of Lavallette History, Lavallette by the Sea*. Borough of Lavallette, 2024. https://www.lavallette.org.

Cartwright, Mark. "Electrical Telegraph." March 24, 2023. *World History Encyclopedia*. https://www.worldhistory.org/Electrical_Telegraph.

CBS 88 News. "November Snowstorm Paralyzes the Tri-State." Aired November 16, 2018.

Cioffi, Joe (meteorologist). "Happy 40th Anniversary, January 9–20, 1978, the Alan Kasper Snowstorm." January 19, 2018. https://www.meteorologistjoecioffi.com.

———. Interview about WPIX 11, New 12 Long Island, WNBC TV, FIOS 1 News, Long Island, Thanksgiving Day snowstorm, November 23, 1989. September 2024.

———. Interview about WPIX 11, WNBC TV, FIOS 1 News, Long Island, News Channel 12 (Long Island), evolution of the Blizzard of '96, synoptic charts, impacts, history. September 2024.

Dabour, Dave (North Jersey Weather Observers). Interview about the great Blizzard of January 22–23, 2016. January 20, 2025.

Dave's Weather. "Major Coastal Blizzard Strikes NYC, NJ On Feb. 11–12, 2006 with Strong Winds, Lightning and Record-Breaking Snowfall… Storm Forms 'Eye' Offshore as Heads for New England." Dave's Weather America. 2024. https://www.facebook.com/people/Daves-Weather-Blog/100063691741500.

Esselin, John. "Boxing Day Snowstorm, December 26, 1947: Loss of Life, People Stranded in New Jesey." NorthJersey.com. January 4, 2018. https://www.northjersey.com/.

Friedman, George. "Memories of Another Bad Winter, the Lindsay Snowstorm, February 9, 1969." *The Jewish Link*, March 6, 2014. https://jewishlink.news/.

Fuller, Betty Ann. Interview about Camp Osborn's origins and history, Blizzard of December 26–27, 2010. March 2024.

Futrell, Jim. "Jenkinson's Pavillion Fire, November 23, 1989, Thanksgiving Eve and Thanksgiving Day Snowstorm." Visit New Jersey. 2004. https://visitnj.org/nj/amusements-activities/amusement-water-parks.

Gaffney Colgan, Kathleen. Interview about the Boxing Day snowstorm, December 26, 1947. October 14, 2024.

Geisenger, Stacy. Interview about the Lindsay Snowstorm.

Goldman, Jeff. "State Police Responded to 1,000 Crashes, Responded to 1,900 Drivers, During NJ Snowstorm." NJ.com. November 19, 2019. https://www.nj.com.

"Governor Christie's Home Lost Power, Snowstorm of October 29, 2011." CBS New York. October 31, 2011. https://www.cbsnews.com.

Gregory, Nick (meteorologist). "Blizzard of February 11–12, 2006 and Christmas Weekend Blizzard of December 26–27, 2010." Fox 5 New York. https://www.fox5ny.com.

Grenoble, Ryan. "Winter Storm Jonas Is Over. Here's the Damage. What's Next?" Huffington Post. January 25, 2016. https://www.huffpost.com.

Grice, Gary K. "The National Weather Service at 150: A Brief History." National Weather Service Heritage. https://vlab.noaa.gov.

Haffert, Pat. "The Great Atlantic Storm of March 5–9, 1962." Sea Isle City Historical Society. https://museumsdatabase.com/. (Storm survivor narrative.)

Hoffman, Bill. "Presidents' Day Blizzard of February 18–19, 2003." *New York Post*, February 18, 2003. https://nypost.com.

Holland, Heather. "Governor Cuomo Orders City Roads and Highways Closed for Blizzard." DNAinfo, New York. December 26, 2015. https://www.dnainfo.com.

Kocin, Paul, and Louis Uccellini. *Northeast Snowstorms*. American Meteorological Society, 2004.

Kosinski, Stan. Interview about Camp Osborn's origins and history. 2024.

Kyle, David (meteorologist). Interview about Blizzard of December 26–27, 2010, Boxing Day blizzard. August 2024.

LaRosa, Joesph A., Jr. *Our Perfect Storm*. Quality Printing, 2010. (Storm survivor narratives.)

Larsen, Eric. "The Great Snowpocalypse of March 1914: 'Asbury Park Cut Off from World.'" *Asbury Park Press*, March 7, 2014.

Leberfinger, Mark. "Blizzard of January 7–8, 1996, Loss of Life, Damages, Impacts." Accu-Weather. January 8, 2016. https://www.accuweather.com/.

Martin, Raymond C. "Snowstorm of February 11–13, 2006. Storm Summary." 2024. https://www.raymondcmartinjr.com.

Mazella, Scott. Interview about the Great Atlantic Storm of 1962, had it occurred in October 2012 instead of Sandy. May 2024.

McCabe, Wayne. Interview about the Blizzard of January 23–24, 1905; and the Blizzard of March 1, 1914. December 10–12, 2024.

McGeehan, Patrick. "Bloomberg Takes Blame for Response to Snowstorm." *New York Times*, December 30, 2010. https://www.nytimes.com.

McQuillan, Allison. "Christmas Weekend Blizzard of December 26–27, 2010." News 4, New York. 2010.

Melisurgo, Len. "Flashback to Brutal Blizzard of January 2016." NJ.com. January 25, 2021. https://www.nj.com.

Mickle, Jean. "Camp Osborn." In *Settlement of Ortley Beach*. Asbury Park Press, 2017.

———. "Learn the Mystery History of Toms River Neighborhoods. Historical Settlement of Ortley Beach, 1812." *Asbury Park Press*, June 22, 2017.

Minutella, Mike, and Sue Minutella. Interview about Camp Osborn's origins and history. 2024.

Mungo, Al (North Jersey Weather Observers). Interview about the Blizzard of February 6–7, 1978; Blizzard of April 6, 1982; megalopolitan Blizzard of February 11–12, 1983; Thanksgiving Day snowstorm of November 23, 1989; "Storm of the Century," March 12–14, 1993; Blizzard of January 7–8, 1996; Blizzard of February 19, 2003; Christmas weekend Blizzard of December 26–27, 2010. 2024.

National Oceanic and Atmospheric Administration. "The Blizzard of March 12–14, Synoptic Chart, U.S Eastern Seaboard." https://www.noaa.gov.

———. "The Storm of the Century: A Look Back from NOAA Satellites, March 12–14, 1993 'Storm of the Century.'" www.NOAA.gov.

Nee, Daniel. "Five Years Later: A Look Back at the Great Brick Blizzard of 2010." *Brick Shorebeat*, December 26, 2015. https://shorebeat.com.

Nesenohm, Gerry Alan. Interview about the Great Blizzard of January 22–23, 2016. September 2024.

News Channel 12, New Jersey. "Blizzard 2006—Night Update, NJ Blizzard of February 11–12, 2006." Aired February 12, 2006.

NJ.com. "Here's What the Snowstorm Looks Like in All 21 Counties of N.J." February 2, 2021. https://www.nj.com/weather/2021/02/heres-what-the-snowstorm-looks-like-in-all-21-counties-of-nj.html.

NOAA, National Weather Service. "Wooly Bear Caterpillar Predicts Early Winter." https://www.weather.gov/arx/woollybear.

Osborn, Jack. Interview about Camp Osborn's history and origins, Blizzard of February 6, 1978. October 2024.

P., Dave. "Blizzard of February 11–12, 2006." Dave's Weather America. 2024. https://www.facebook.com/people/Daves-Weather-Blog/100063691741500.

Pharo, Jill. Interview about the Great Atlantic Storm of March 5–9, 1962, in Beach Haven, New Jersey. (Storm survivor narrative.)

"Powerful Blizzard Heading North." NBC News, Philadelphia. Aired December 26, 2010.

"Presidents' Day Blizzard of February 17–18, 2003." CNN. 2003. https://www.cnn.com.

Rao, Joe (meteorologist). Interview about the Lindsay snowstorm, February 9, 1969. June 28, 2025.

———. "Megalopolitan Blizzard: WABC Air Clip." WABC News Radio. Aired February 11, 1983.

———. "The Storm of the Century, March 12–14, 1993." News Channel 12. Aired March 12, 1993.

Robinson, David (New Jersey state climatologist). "The Blizzard of January 22–23, 2016." Rutgers–New Brunswick: School of Arts and Sciences. https://geography.rutgers.edu.

Salvini, Emil R. "The Great Atlantic Storm of 1962." NJ Spotlight News, powered by NJPBS. https://www.njspotlightnews.org.

"Seaside Heights Boardwalk Destruction, NJ Storm Damages: Blizzard of February 6–7, 1978." *Newark Star Ledger*, February 9, 2013. https://www.nj.com/news/2013/02.

Smith, Wayne. Interview about Camp Osborn's origins and history, 1938 hurricane. 2024.

"Snowfall Rate Climbs Into 30-Inch Range in Several Union County Towns." *Star Ledger*, January 26–27, 2010.

"Snowpacalypse Now." *New York Post*, December 27, 2010. https://nypost.com.

Stacheleski, Christopher (meteorologist). "The Blizzard of '96: In Retrospect, 25 Years Later." National Weather Service Heritage. 2021. https://vlab.noaa.gov.

Stefano, Nick (North Jersey Weather Observers). Interview about the great Blizzard of January 22–23, 2016; the Blizzard of January 22–23, 2010. January 14, 2025.

Thomas, Eric (meteorologist for WBTV in Charlotte, NC). "The Storm—Not of the Century—In the History of Man-Kind, the Most Powerful Storm to Ever Affect the East Coast of the United States!" www.NOAA.gov.

Thomas Buchholz, Margaret. Interview about the Great Atlantic Storm of March 5–9, 1962, her storm story. October 2024.

Thomas Buchholz, Margaret, and Raymond C. Fisk. *Great Storms of the Jersey Shore*. Down the Shore Publishing, 2024.

Tursi, Frank. "Great Atlantic Storm of March 5–9, 1962: Damage to Cape Hatteras Outer Banks." CoastalReview.org. March 9, 2012. https://coastalreview.org.

UPI Archives. "April Blizzard Slams New York City." April 6, 1982.

Valle, Dan. "Blizzard of March 12–14, 1888." National Weather Service Heritage. 2024. https://vlab.noaa.gov.

Weather Underground. "The Great Blizzard of January 22–23, 2016, Newark, New Jersey Weather History." https://www.wunderground.com.

Witt, Jim (meteorologist). Interview about the Boxing Day snowstorm, December 26, 1947. June 2024.

"You Think This Storm Is Bad? 1978's Blizzard Crippled N.J. For Days." Newark Star Ledger Staff. February 9, 2013.

Zimmer, David. "Boxing Day Snowstorm, December 26, 1947: Snowfall Accumulations in New Jersey, NYC." December 16, 1922. NorthJersey.com. https://www.northjersey.com.

ABOUT THE AUTHOR

Don Colgan has nurtured a lifetime passion in meteorology, particularly winter storms. A lifelong resident of Central New Jersey, he has written recreational fishing articles and several children's short stories. Recently retired from a forty-eight-year career in sales and marketing in the field of electrical distribution, Colgan loves storytelling, and his lifelong ambition was to author a book about the historic winter storms that have punished the Garden State through the eyes and words of New Jerseyans who lived through them.